TRANSACTIONS

OF THE

AMERICAN PHILOSOPHICAL SOCIETY

HELD AT PHILADELPHIA
FOR PROMOTING USEFUL KNOWLEDGE

NEW SERIES—VOLUME 58, PART 3
1968

SANSKRIT ASTRONOMICAL TABLES IN THE UNITED STATES

DAVID PINGREE
Oriental Institute, University of Chicago

THE AMERICAN PHILOSOPHICAL SOCIETY
INDEPENDENCE SQUARE
PHILADELPHIA

APRIL, 1968

Library of Congress Catalog
Card Number 68-19148

PREFACE

This catalogue is an attempt to organize and put into a somewhat comprehensible state an enormous mass of information gleaned during many delightful weeks of reading Sanskrit manuscripts in Cambridge, New York, and Philadelphia during the summers of 1963 and 1966. The author's sincerest thanks are proffered to the custodians of the Harvard, Columbia, and University of Pennsylvania Libraries for making this search through a small part of their rich collections possible.

The manuscripts examined for inclusion in this catalogue are those which contain primarily astronomical tables; any texts that they contain should ordinarily be there only to explain the use of the tables. Thus all the common siddhântas, karaṇas, and works on instruments were excluded from consideration, though manuscripts of such texts normally contain tables. A few special treatises were left out as well; these include the *Keśavīpaddhati* of Keśava, which is mainly astrological and which has, in any case, been published several times, and the *Pâtagaṇita* of Gaṇeśa, which I am still far from understanding completely. Beyond these limits it is hoped that all relevant manuscripts have been examined, and that the functions of most of the tables investigated have been properly comprehended.

The presentation of this type of material presents many problems. For the purpose of removing from the reader's path some of the difficulties that he may encounter in trying to use the catalogue it may be useful here to outline the arrangement that has been followed. The introduction discusses the origins and dates of the three main collections from which the manuscripts are drawn so that one may have a clearer idea of just what the material represents in terms of the whole corpus of Sanskrit astronomical tables. Incidental to establishing provenience, it was felt that lists of the manuscripts of the three collections—considerably more correct than the entries in Poleman's *Census*—would be useful; anyone familiar with Sanskrit manuscripts of texts on jyotiḥśâstra will immediately realize whence these collections come.

Following the introduction is the catalogue of manuscripts proper, in which the entries are arranged in the order of their sequence in Poleman's *Census*. In describing each manuscript information has first been given concerning its extent, its date, and its scribe, and then, folio by folio, an indication of its contents; here reference is constantly made to the descriptions of texts and tables in the third part of this volume. Data concerning the measurements of the manuscripts can be found in Poleman and are not repeated here.

The final section contains detailed analyses of the various works contained in the manuscripts examined for this catalogue. These works are dealt with in the order of their dates of composition in so far as these dates can be determined; the several sets of tables for which no dates are available are gathered together at the end. In each case an attempt has been made to assemble reliable information concerning the author, when he is identified, and a list of all manuscripts of the work which are at present known to me is also provided; further details in both of these categories will in several cases be given in my *Census of the Exact Sciences in Sanskrit,* now in the course of preparation. The descriptions of the tables themselves attempt to make three things clear: the function of the tables—that is, what new information they furnish and what prior information (argument) is necessary in order to elicit this new information; their structure—that is, how they are arranged to make the elicitation of this new information possible; and their parameters—that is, the quantitative basis of the new information. A few tables have been transcribed from the manuscripts as illustrations.

In most cases no attempt has been made to discover the method used by the Indian astronomer in computing the tabulated entries, and in a few instances (notably with regard to some of the eclipse-tables) my statements concerning functions are not to be considered definitive. But it is hoped that sufficient information has been provided so that historians of astronomy will know the types of tables available in Western and Northern India during the half-millennium which preceded the introduction of a European educational system, and will better understand what problems Indian astronomers were interested in solving and how they transformed their solutions into tables; and so that cataloguers of Sanskrit manuscripts will no longer be able to pass off these massive products of Indian science as unidentifiable sâraṇîs. Hopefully the same service can eventually be performed for manuscript collections representing Eastern and (most importantly) Southern India.

SANSKRIT ASTRONOMICAL TABLES IN THE UNITED STATES

David Pingree

CONTENTS

INTRODUCTION

This catalogue of Sanskrit astronomical tables in American libraries is an initial effort to deal with the immense corpus of such materials preserved in manuscript collections throughout the world. The purpose is to indicate the structure of the tables, to determine (wherever possible) their functions, and to extract from them significant parameters. It is hoped that this analysis will assist in the identification of other manuscript materials as well as provide information concerning a seldom explored field in the history of astronomy.

The foundation upon which the present catalogue rests is H. Poleman's *A Census of Indic Manuscripts in the United States and Canada* (*Amer. Orient. Soc.* 12, New Haven 1938). It was, of course, beyond the scope of that work to undertake an investigation like the present one; but without Poleman's preliminary listing of tables the task of preparing this catalogue would have been much more difficult. Therefore, despite the many corrections which this catalogue makes in Poleman's identifications and classifications, it was felt to be appropriate to retain his numbers as the primary reference system.

Manuscripts of Sanskrit astronomical tables are preserved in three libraries in this country: that of Columbia University in New York, that of Harvard University in Cambridge, and that of the University of Pennsylvania in Philadelphia. The Sanskrit manuscripts preserved at Columbia were purchased by David Eugene Smith. The Indic collection (included in Poleman) consists of more than 248 items, of which the first 194 are apparently from Gujarat and Rajasthan, or perhaps Uttar Pradesh, and the remainder mostly from Ceylon, Burma, Thailand, Cambodia, and Indonesia. The Sanskrit collection (not included in Poleman) consists of 38 manuscripts from Western India. The Harvard collection comes from many sources; those noted here are from Western India also. Another Harvard collection of 11 manuscripts (not included in Poleman) consists of items from Bengal. The University of Pennsylvania collection was gathered by several people, though two substantial collections (constituted in India or in Philadelphia?) make up its core. These manuscripts again are almost entirely from Uttar Pradesh, Gujarat, and Rajasthan. This catalogue, then, represents the material which was current in Northern and Western India in the last century.

Herewith is a list of manuscripts 1 to 194 in the Smith Indic collection. The Poleman numbers are given in parentheses. Those manuscripts which are marked with an asterisk have been inspected by the present writer.

1. (4762) *Līlāvatīvṛtti*, a commentary on the *Līlāvatī* of Bhāskara II. 10 ff.
2. (4747 and 4761) *Līlāvatī* of Bhāskara II with a commentary by Sūrya. 12 ff. and ff. 2–44.
3. (4750) *Līlāvatī* of Bhāskara II. 46 ff. Copied by Mallāri in Śaka 1620 (A.D. 1698).

4. (4751) *Līlāvatī* of Bhāskara II. ff. 1 and 36–59.
*5. (4752) *Līlāvatī* of Bhāskara II. 54 ff. Copied by Karuṇāśaṅkara the son of Śivaśaṅkara in Saṃ. 1842, Śaka 1707 (A.D. 1785).
6. (4753) *Līlāvatī* of Bhāskara II. ff. 1–18 and 11–32, and 1 unnumbered f. Copied by Bhaṭṭarāmeśvara in Saṃ. 1774, Śaka 1639 (A.D. 1717).
7. and 8. (4741) *Vāsanābhāṣya* of Bhāskara II, a commentary on his own *Siddhāntaśiromaṇi*. 137 ff.
9. (4738) *Siddhāntaśiromaṇi* of Bhāskara II. 9 ff.
10. (4743) *Gaṇitādhyāya* of Bhāskara II. 42 ff. Copied by Nandarāma at Benares in Saṃ. 1882 (A.D. 1825).
11. (4745) *Golādhyāya* of Bhāskara II. ff. 1–5, 7, and 14–18.
12. (4754) *Līlāvatī* of Bhāskara II. ff. 1–20a, 20b–38, and 1–9.
13. (4755) *Līlāvatī* of Bhāskara II. ff. 1–4, 6–20a, and 20b–47. Copied in Saṃ. 1948, Śaka 1813 (A.D. 1891).
14. (5077) *Bṛhatsaṃhitāvivṛti*, a commentary by Utpala on the *Bṛhatsaṃhitā* of Varāhamihira. ff. 1–302, 304–614, 140, 615–807, 809–871, 873–976, and 978–1114.
*15. (——) *Laghujātaka* of Varāhamihira with a commentary by Utpala. 44 ff. Copied by Nandarāma in Saṃ. 1883 (A.D. 1826).
*16. (4708) *Bṛhattithicintāmaṇi* of Gaṇeśa. 39 ff.
*17. (4950) *Khecaradīpikā* of Kalyāṇa. 12 ff.
*18. (4767) *Mahādevī* of Mahādeva. ff. 1 and 3–7.
*19. (4949) *Candrārkī* of Dinakara. 2 ff.
*20. (5224) Tables of daśās, antardaśās, and upadaśās. 7 ff.
*21. (4878) Anonymous of 1714. 2 ff.
*22. (4765) *Mahādevī* of Mahādeva. 3 ff.
*23. (4894) Two dinamānapattras. 2 ff.
24. (5967) *Rāhusādhanapratīkāra*; in Hindi. 4 ff.
*25. (4917) Anonymous II of 1578. 1 f.
*26. (4694) *Grahalāghavasāriṇī*. 6 ff.
*27. (4807) Anonymous of 1598. 1 f. Copied by Malūkacandra at Rādhanapura (Radhanpur, W.I.).
*28. (5177) Anonymous of 1741. ff. 8–23. Copied by Amarasiṃha in Saṃ. 1846 (A.D. 1789).
*29. (4952) *Candrārkī* of Dinakara and *Brahmatulyasāraṇī*. 6 ff. Copied by Malūkacandra.
*30. (4778) *Pancāṅgavidyādharī* of Vidyādhara. 18 ff. Copied by Josī Hīrajī at Māṇḍavi (Mandvi, Cutch, W.I.) in Saṃ. 1806 (A.D. 1749).
31. (5213) *Janmapattrikā*. 8 ff.
32. (5069) *Bṛhajjātakavivaraṇa* of Mahīdhara. ff. 1–4 and 6–63. Copied in Saṃ. 1856, Śaka 1721 (A.D. 1799).
*33. (4777) *Grahaprabodhasāriṇī* of Yādava. 34 ff. Copied by Govindaśarman Nāsikara in Śaka 1793 (A.D. 1871).
*34. (4883) *Candrārkī* of Dinakara. ff. 9–11.
*35. (5178) *Candrārkī* of Dinakara. 11 ff.
36. (4981) *Jātakābharaṇa* of Ḍhuṇḍhirāja. 101 ff. Saṃ. 1863, Śaka 1728 (A.D. 1806).
37. (5068) *Bṛhajjātakavivṛti* of Utpala. ff. 1–274a, 274b–284, 286–306a, 306b–315, 317–378. Saṃ. 1847 (A.D. 1790).
38. (5071) *Bṛhatsaṃhitā* of Varāhamihira. 104 ff. and ff. 1–39, 44–66.
39. (4784) *Bhāsvatyudāharaṇa*. 10 ff.
*40. (4895) *Candrārkī* of Dinakara. 4 ff.
*41. (4763a) *Mahādevī* of Mahādeva and Anonymous III of 1578. 128 ff. Copied by Viṭṭhalajī in Śaka 1588 (A.D. 1666).
*42. (5200) *Prauḍhamanoramā*, a commentary on Keśava's *Keśavīpaddhati* by Divākara. 177 ff. Copied by Bhagavanta Daivajña in Śaka 1704 (A.D. 1782).
*43. (4876) *Brahmatulyasāraṇī*. 13 ff.
44. (5025) *Ṣaṭpañcāśikā* of Pṛthuyaśas with a commentary. 22 ff. Saṃ. 1860 (A.D. 1803).
*45. (4735) *Brahmatulyasāraṇī*. ff. 2–10 and 12–17. Saṃ. 1855, Śaka 1720 (A.D. 1798). f. 11 equals 4946 LXII.
*46. (5179) *Candrārkī* of Dinakara. ff. 2–10.
47. (5054) *Muhūrtacintāmaṇiṭīkā*, a commentary by Rāma on his own *Muhūrtacintāmaṇi*. 40 ff.
48. (5199) *Makarandodāharaṇa*, a commentary by Viśvanātha on Makaranda's *Makaranda*. 28 ff. Saṃ. 1832 (A.D. 1775).
*49. (4720) *Makarandavivaraṇa*, a commentary by Divākara on Makaranda's *Makaranda*. 19 ff. Saṃ. 1922 (A.D. 1865).
50. (4982) *Jātakābharaṇa* of Ḍhuṇḍhirāja. 110 ff. Copied by Vidyādhara in Saṃ. 1918 (A.D. 1861).
51. (4732) *Brahmatulyodāharaṇa*, a commentary by Viśvanātha on the *Brahmatulya* of Bhāskara II. 16 ff.
*52. (4924) Anonymous of 1704. 2 ff.
*53. (4718) *Tithyādicintāmaṇi* of Dinakara. 2 ff.
54. (5059) *Muhūrtadīpaka* of Rāmasevaka. 17 ff. Saṃ. 1836 (A.D. 1779).
55. (5247) *Kapardikāpraśna*. 6 ff. Saṃ. 1866 (A.D. 1809).
56. (5098) *Āyuḥpraśna*. 4 ff. Saṃ. 1864, Śaka 1729 (A.D. 1807).
57. (4837) *Jñānamañjarī*. 14 ff.
*58. (4825) *Candrārkī* of Dinakara. 15 ff.
59. (4723a) *Ariṣṭanavanītaṭīkā*, a commentary by Śrīdhara on Navanīta Kavi's *Ariṣṭanavanīta*. 7 ff. Saṃ. 1874 (A.D. 1817).
60. (5194) *Jātakapaddhatyudāharaṇa*, a commentary by Kṛṣṇa on Keśava's *Keśavīpaddhati*. 109 ff.
61. (4991) *Muhūrtamārtaṇḍa* of Nārāyaṇa. 32 ff. Saṃ. 1889 (A.D. 1832).
62. (5056) *Muhūrtacintāmaṇiṭīkā*, a commentary on Rāma's *Muhūrtacintāmaṇi*. 5 ff.
63. (5245) *Mayūracitraka*, chapter 46 of the *Bṛhat-*

saṃhitā of Varāhamihira. 9 ff. Copied by Vidyādhara in Saṃ. 1867 (A.D. 1810).
64. (4797) *Arghakāṇḍa* from the *Trailokyaprakāśa* of Hemaprabha Sūri. 13 ff. Saṃ. 1904, Śaka 1769 (A.D. 1847).
65. (5060) *Muhūrtabhūṣaṇa* (*Muhūrtadīpaka*) of Rāmasevaka. 13 ff. Copied by Vidyādhara in Saṃ. 1865, Śaka 1730 (A.D. 1808).
66. (4920) *Saṃgrahāvalī*. 2 ff.
67. (5058) *Muhūrtadīpaka* of Rāmasevaka. ff. 8–21. Saṃ. 180– (A.D. 1743–1752).
68. (5057) *Muhūrtadīpaka* of Rāmasevaka. 10 ff.
69. (5089) *Pañcāśapraśna* of Śaṅkara. 3 ff. Saṃ. 1846 (A.D. 1789).
70. (4591) *Yoginīdaśā* (from the *Rudrayāmala?*). 8 ff. Copied by Rāmaphalarāma in Saṃ. 1872 (A.D. 1815).
71. (5016) *Pārāśarī* of Parāśara. 9 ff. Copied by Cītharaśarman Miśra in Saṃ. 1868 (A.D. 1811).
72. (4836) *Jñānapradīpa*. 2 ff.
73. (4715) *Yantrarāja* of Jayasiṃha. 35 ff.
74. (4712) *Yantradīpikā*, a commentary by Rāma on Cakradhara's *Yantracintāmaṇi*. 21 ff. Saṃ. 1929 (A.D. 1872).
*75. (4937) Table. 1 f.
76. (5051) *Muhūrtacintāmaṇiṭīkā*, a commentary (by Kāśīnātha?) on Rāma's *Muhūrtacintāmaṇi*. ff. 6–47 and 49–50.
77. (5223) Horoscope. 1 f. Saṃ. 1900, Śaka 1765 (A.D. 1843).
78. (5017) *Jātakacandrikā* of Parāśara. 2 ff. Saṃ. 1889 (A.D. 1832).
*79. (4721) *Makarandavivaraṇa*, a commentary by Divākara on Makaranda's *Makaranda*. ff. 6–7. Saṃ. 1853, Śaka 1718 (A.D. 1796).
*80. (4764) *Mahādevī* of Mahādeva. 9 ff.
*81. (4863) Anonymous of 1578. 3 ff.
*82. (5180) Anonymous of 1741. 3 ff.
83. (5101) *Kāmadhenu*. 16 ff.
*84. (4864) Tables based on the *Mahādevī*. 2 ff.
*85. (4865) Tables based on the *Mahādevī*. 2 ff.
*86. (4887) Māsapraveśapattra. 1 f.
*87. (5181) Nakṣatracaraṇapraveśapattra. 1 f.
88. (4690) *Grahalāghava* of Gaṇeśa. 12 ff.
89. (4992) *Muhūrtamārtaṇḍa* of Nārāyaṇa. ff. 1–4, 7–10, 12–16, and 22.
90. (5219) *Mahādaśāpattra*. 1 f. Saṃ. 1945, Śaka 1810 (A.D. 1888).
*91. (4771) *Mahādevī* of Mahādeva. 8 ff.
*92. (4680) *Tithicintāmaṇi* and *Bṛhattithicintāmaṇi* of Gaṇeśa. 12 ff. Śaka 1704 (A.D. 1782).
93. (5260) *Pallīphalavicāra*. 1 f.
*94. (4679, 4678, 4681, and 4682) *Tithicintāmaṇi* of Gaṇeśa. 40 ff. Śaka 1652 (A.D. 1730).
*95. (4858) Dinamānapattra. 1 f. Saṃ. 1916, Śaka 1781 (A.D. 1859).
*96. (4903) Bhāvapattra. 2 ff.
97. (5095) Antardaśācakra. 3 ff.
*98. (4886) Māsapraveśapattra. 1 f. Saṃ. 1861, Śaka 1727 (A.D. 1804/5).
99. (4871) Pañcakavargas. 8 ff.
*100 (4805, 4821, 4822, 4884, 4904, and 4916). Anonymous of 1704. ff. 3–22.
*101. (4866) *Mahādevī* of Mahādeva. 8 ff.
102. (5249) *Lokamanoramā* of Garga with a commentary. 5 ff. Saṃ. 1881 (A.D. 1824).
103. (5350) *Vāstupradīpa* of Vāsudeva. 10 ff. Saṃ. 1882 (A.D. 1825).
104. (4719) *Divākarīyapaddhati* of Divākara. 5 ff. Saṃ. 1890, Śaka 1755 (A.D. 1833).
*105. (4867) *Mahādevī* of Mahādeva. 46 ff.
106. (5065) *Bṛhajjātaka* of Varāhamihira. ff. 1–14, 17–40, and 42–44. Copied by Devīdatta Tripāṭhin in Saṃ. 1874, Śaka 1739 (A.D. 1817).
107. (4938) *Pañcāṅga* for Saṃ. 1821, Śaka 1686 (A.D. 1764).
108. (5232) *Lampāka* of Padmanābha. 8 ff. Saṃ. 1897 (A.D. 1840).
109. (4774) *Yantrarāja* of Mahendra Sūri. 15 ff. Copied by Jītaśekhara in Saṃ. 1822 (A.D. 1765).
110. (4776) *Yantrarājaṭīkā*, a commentary by Malayendu Sūri on the *Yantrarāja* of Mahendra Sūri. 28 ff. Copied by Śaṅkara, the son of Śrīdhara, in Saṃ. 1780 (A.D. 1723).
111. (4782) *Bhāsvatī* of Śatānanda. 10 ff. Saṃ. 1879 (A.D. 1822).
112. (4926) *Sūryasiddhānta*. 15 ff.
*113. (4889) Dinamānapattra. 2 ff.
114. (4912) *Vāstusuptacandracakra*. 9 ff.
115. (5267) *Meghamālā* ascribed to Rudra (Śiva). 20 ff.
116. (4941) *Pañcāṅga* for Saṃ. 1910, Śaka 1775 (A.D. 1853). 13 ff.
117. (5251) *Pañcapakṣī*. 11 ff.
118. (5252) *Pañcapakṣīpraśna*. 4 ff.
*119. (4942) Anonymous of 1594 and Anonymous of 1638. ff. 2–18.
120. (5166) *Śukradaśā*. 4 ff.
121. (4909) *Varṣaphaladṛṣṭivicāra* (?). 2 ff.
122. (5268) *Ramalanavaratna* of Paramasukha. 16 ff. Saṃ. 1892 (A.D. 1835).
123. (5072) *Bṛhatsaṃhitā* of Varāhamihira. 82 ff., 32 ff., and ff. 36–54.
124. (5275) *Pañcāṅga* for Saṃ. 1875, Śaka 1740 (A.D. 1818). 15 ff.
125. (4794) *Jātakālaṅkārakarman* of Śrīśuka. 5 ff.
*126. (4846) *Jyotiṣasiddhāntasaṅgraha*, for a nativity in Saṃ. 1871, Śaka 1736 (A.D. 1814). 16 ff.
127. (4723) *Yantrasāra* of Nandarāma. 27 ff. Copied by Rāmakṛṣṇa in Saṃ. 1891 (A.D. 1834).
128. (5176) *Hillājadīpikā*. 12 ff.
*129. (4800, 4801, 4802, 5187, and 5188). *Tithicintāmaṇi* of Gaṇeśa and *Makaranda* of Makaranda. 33 ff., ff. 2–21, and 8 ff.

130. (5012a) *Nīlakaṇṭhyudāharaṇa*, a commentary by Viśvanātha on Nīlakaṇṭha's *Nīlakaṇṭhī*; this manuscript contains only the *Saṃjñātantra* and the *Varṣatantra*. 94 ff.
131. (5036) *Bālabodha* of Muñjāditya. ff. 9–52 and 27–29. Copied by Raghunātha in Saṃ. 1721 (A.D. 1664), Śaka 1568 (A.D. 1646). Read Śaka 1586 (A.D. 1664).
132. (5003) *Nīlakaṇṭhī* of Nīlakaṇṭha; this manuscript contains the *Varṣatantra* only. 44 ff. Śaka 1667 (A.D. 1745).
133. (4691) *Grahalāghava* of Gaṇeśa. 15 ff.
134. (4983) *Bhuvanadīpaka*. ff. 1–23a and 23b–25. Saṃ. 1877 (A.D. 1820).
135. (5269) *Ramalanavaratna* of Paramasukha. ff. 1–12, 11, and 14–53.
136. (4731) *Brahmatulyodāharaṇa*, a commentary by Viśvanātha on the *Brahmatulya* of Bhāskara II. ff. 1–30, 34, 33, 31–32, and 35–108. Saṃ. 1853, Śaka 1718 (A.D. 1796).
137. (4971) *Yuddhajayotsava* of Gaṅgārāma. ff. 2–6 and 8–24.
*138. (4855, 4861, 4892, 4859, 5125, 4951, and 4943) *Tithicintāmaṇi* of Gaṇeśa and Anonymous of 1461. 28 ff.
*139. (4766) Anonymous of 1704. 138 ff.
*140. (4868) *Mahādevī* of Mahādeva and Anonymous III of 1578. 177 ff.
141. (4677) *Varṣaphalapaddhatiṭīkā*, a commentary by Mallāri on a part of Keśava's *Keśavīpaddhati*. 14 ff.
142. (4695) *Siddhāntarahasyodāharaṇa*, a commentary by Viśvanātha on Gaṇeśa's *Grahalāghava*. 56 ff.
143. (4783) *Bhāsvatīṭīkā*, a commentary by Rāmakṛṣṇa on Śatānanda's *Bhāsvatī*. ff. 3–24.
144. (4781) *Karaṇavaiṣṇava* of Śaṅkara. ff. 1–23 and 28–51.
145. (5198) *Keśavīpaddhatyudāharaṇa*, a commentary by Viśvanātha on Keśava's *Keśavīpaddhati*. 77 ff. Copied by Śivabhaṭa, the son of Sadāśivabhaṭa, in Śaka 1713 (A.D. 1791).
*146. (4869) *Jagadbhūṣaṇa* of Haridatta. 100 ff.
*147. (4769 and 4773) *Mahādevī* of Mahādeva. 61 ff.
148. (5070) *Bṛhajjātakavivaraṇa*, a commentary by Mahīdhara on Varāhamihira's *Bṛhajjātaka*. 36 ff.
149. (5241) *Samarasāra* of Rāmacandra. 10 ff.
150. (5088) *Camatkāracintāmaṇi* of Vrajabhūṣaṇa. 11 ff. Saṃ. 1875 (A.D. 1818).
*151. (4709) *Bṛhattithicintāmaṇi* of Gaṇeśa. 107 ff. Śaka 1682 (A.D. 1761).
*152. (4902) *Rāhucāra*, chapter 5 of Varāhamihira's *Bṛhatsaṃhitā*. 10 ff.
153. (4813) *Keśavīpaddhati* of Keśava. 21 ff.
154. (5048) *Muhūrtacintāmaṇi* of Rāma. 50 ff. Saṃ. 1904 (A.D. 1847).
155. (5190) *Keśavīpaddhati* of Keśava. 8 ff. Saṃ. 1848 (A.D. 1791).
156. (5191) *Keśavīpaddhati* of Keśava. 6 ff. Saṃ. 1884, Śaka 1749 (A.D. 1827).
157. (4975) *Praśnavidyā* (*Lokamanoramā*) of Garga with a commentary. 6 ff. Saṃ. 1925 (A.D. 1868).
158. (5129) *Padmakośa*. 10 ff. Saṃ. 1907, Śaka 1772 (A.D. 1850).
159. (5049) *Muhūrtacintāmaṇi* of Rāma with the author's own commentary. 44 ff.
160. (5004, 5008, and 5002) *Nīlakaṇṭhī* of Nīlakaṇṭha. 42 ff. Copied by Rāmadina Miśra in Saṃ. 1885, Śaka 1750 (A.D. 1828).
161. (5024) *Ṣaṭpañcāśikā* of Pṛthuyaśas with a commentary. ff. 1–9a, 9b, 11–19a, and 19b–20. Copied by Harinandana in Saṃ. 1881, Śaka 1745 (A.D. 1823/4).
162. (4993) *Muhūrtamārtaṇḍa* of Nārāyaṇa. 19 ff. Copied by Kāśīnātha in Śaka 1722 (A.D. 1800).
163. (5040) *Muhūrtamañjarī* of Yadunandana. ff. 6–12.
164. (4737) *Siddhāntaśiromaṇi* of Bhāskara II. ff. 57–70 and 72–118.
165. (5243) *Samarasāraṭīkā*, a commentary on Rāmacandra's *Samarasāra*. 13 ff.
*166. (4853, 4857, and 4862) *Makaranda* of Makaranda. 31 ff.
167. (4697) *Siddhāntarahasyodāharaṇa*, a commentary by Viśvanātha on Gaṇeśa's *Grahalāghava*. 82 ff. Copied by Rāmakṛṣṇa in Śaka 1704 (A.D. 1782).
168. (4891) *Yantrarājaracanāprakaraṇa* (of Jayasiṃha?). 3 ff.
169. (4733) *Brahmatulyodāharaṇa*, a commentary by Viśvanātha on the *Brahmatulya* of Bhāskara II. 63 ff.
170. (4832) *Upadeśasūtra* of Jaimini. 10 ff.
171. (4833) *Upadeśasūtra* of Jaimini. 8 ff. Saṃ. 1928 (A.D. 1871).
*172. (4944) *Mahādevī* of Mahādeva. 151 ff. Copied by Amarasiṃha in Saṃ. 1814 (A.D. 1758).
*173. (4879) *Makarandavivaraṇa*, a commentary by Divākara on Makaranda's *Makaranda*. 5 ff.
174. (4711) *Makarandodāharaṇa*, a commentary by Viśvanātha on *Makaranda's* Makaranda. 12 ff. Saṃ. 1881 (A.D. 1824).
175. (4692) *Grahalāghava* of Gaṇeśa. 24 ff. Copied by Badalarāma Tripāṭhin.
176. (4696) *Siddhāntarahasyodāharaṇa*, a commentary by Viśvanātha on Gaṇeśa's *Grahalāghava*. ff. 16–30.
177. (4683) *Grahalāghava* of Gaṇeśa. 9 ff.
178. (4684) *Grahalāghava* of Gaṇeśa, 19 ff. Saṃ. 1901 (A.D. 1844).
*179. (4770) *Mahādevī* of Mahādeva and the Anonymous of 1741. ff. 3–16.

*180. (4827) *Candrārkī* of Dinakara. 2 ff. Copied by Malūkacandra in Saṃ. 1829, Śaka 1694 (A.D. 1772) at Rādhanapura (Radhanpur, W.I.).
181. (4939) A pañcāṅga. 16 ff.
182. (4940) *Pañcāṅga* for Saṃ. 1922, Śaka 1787 (A.D. 1865). 31 ff.
183. (4960) *Praśnapradīpa* of Kāśīnātha. 6 ff.
·184. (5099) *Āyurdāya*. 9 ff.
*185. (4877 and 4675) *Makaranda* of Makaranda and *Śīghrabodha* of Kāśīnātha. 55 ff.
186. (5170) *Saṃkrāntiphala*. 2 ff.
*187. (4922) On the correspondences between akṣaras and astrological places, zodiacal signs, nakṣatras, elements, etc. 8 ff.
188. (5186) Unidentified astrological text. 2 ff.
189. (5110 and 4803) *Ghātacakranirṇaya* and *Andhāvandhākhyavicāra*. 2 ff.
*190. (4717, 4923, 4824, and 4823) *Candrārkī* of Dinakara. ff. 3–6 and 8–17.
*191. (4870) Tables based on the *Mahādevī* of Mahādeva. 2 ff.
*192. (4763) *Mahādevī* of Mahādeva. 29 ff.
*193. (4772) *Mahādevī* of Mahādeva. ff. 49–75.
*194. (4945) *Tithicintāmaṇi* of Gaṇeśa, Dinamānapattra, Anonymous of 1594, and *Makaranda* of Makaranda. 18 ff.

With this group also belongs the so-called *Miscellaneous Bundle (4946), the contents of which will be described in the following pages. Of all these manuscripts, the only one which is at all unusual in a collection formed in Western India is 59 (4723a), the *Ariṣṭanavanīta* of Navanīta Kavi; but this work is sometimes found in the area (e.g., N-W. Prov. IX (1885) 5 and PL, Buhler 6; the four Baroda manuscripts seem to be of Southern origin). At least two of the manuscripts—42 (5200) and 151 (4709)—are still held in the wrappers of a Bombay book-dealer, Pandit Jyeshtaram Mukundji. The Pandit is perhaps the author of the estimates of length (śloka —) which appear on many of the manuscripts and were presumably used to determine the price. The European numerals are probably shelfmarks which the manuscripts had in some library.

The remaining manuscripts in the Smith Indic collection are almost all Buddhist texts from Southeast Asia written in local scripts and in Pali or in the local language. Though there are some jyotiṣa manuscripts in these local languages—e.g., 195, 198–200, and 211 (app. p. 403) in Burmese; 196 (7146), 202 (7148), 212 (7145), and 245 (7149), in Sinhalese; and 197 (5554), 215 (5550), and 216 (5551) in Burmese script—the only Sanskrit texts are the following:

210. (4931) *Sūryasiddhāntodāharaṇa* a commentary by Nṛsiṃha on the *Sūryasiddhānta*. 22 ff.
214. (4668) *Āryabhaṭīya* of Āryabhata I. 70 ff. In Grantha script.
241. (4739) *Siddhāntaśiromaṇi* of Bhāskara II. 13 ff.
242. (4759) *Līlāvatī* of Bhāskara II with a commentary. 88 ff. In Malayalam script.
243. (4929) *Sūryasiddhānta* with a commentary. 78 ff.

Presumably all five of these manuscripts come from South India. One has only to add that two numbers of Smith Indic are not found in Poleman: *221 and 247. The latter is presumably 6543, and the former is a Cambodian manuscript of 29 ff. (ff. 1–5 are blank) purchased in Bangkok in 1930. My ignorance of the Cambodian script prevents any further identification.

The collection of 38 Smith Sanskrit manuscripts contains little of interest to the history of jyotiḥśāstra; I note the following from the handwritten list at Columbia University:

3. *Tithinirṇaya* of Rāghava. A.D. 1768.
4. *Kālasiddhāntanirṇaya* of Candracūḍa.
*5. *Cintāmaṇiṭippaṇa*, a commentary by Vyeṅkaṭa on Gaṇeśa's *Tithicintāmaṇi*. 24 ff.
11. *Svapnādhyāya*.
12. *Pañcāṅgasādhana* (*Tithicintāmaṇi* of Gaṇeśa?)
19. *Svapnādhyāya*. A.D. 1735.
26. *Narapatijayacaryā* of Narapati. A.D. 1705.
36. *Kālanirṇayadīpikā* of Rāmacandrācārya.

As noted above, there are two main series of jyotiṣa manuscripts in the University of Pennsylvania collection. The first of these comprises numbers 648 to 713; its original labels bear numbers in (two) series entitled *Jyotisham*, written in the Roman alphabet. The second comprises numbers 1766 to 1916; its labels bear numbers in a series entitled *Jyotiṣavibhāga*, written in the Devanāgarī alphabet. Both series are constituted of manuscripts from Western, Northern, and Central India; I append a list of each. The first number is the current University of Pennsylvania number; the second the number of the manuscript in the *Jyotisham* or *Jyotiṣavibhāga* series; and the third the Poleman number.

*648 (1) (4841) *Jyotiṣavedāṅga* of Lagadha. 4 ff.
649 ( ) (4756) *Līlāvatī* of Bhāskara II. 14 ff.
650 ( ) (5037) *Bālabodha* of Muñjāditya. 21 ff. Saṃ. 1867 (A.D. 1810).
651 ( ) (5006) *Nīlakaṇṭhī* of Nīlakaṇṭha; this manuscript contains only the *Varṣatantra*. ff. 5–18.
*652 (1!) (4394) *Tattvārthacintāmaṇi* of Vedavṛkṣa Bālakṛṣṇa with the author's own commentary. 7 ff.
653 (2) (5277) *Sarvārthacintāmaṇi*. ff. 1 and 3–60.
654 (3) (5012c) *Nīlakaṇṭhyudāharaṇa* (?), a commentary by Viśvanātha on Nīlakaṇṭha's *Nīlakaṇṭhī*. 23 ff.
655 (4) (5093) *Saṅketakaumudī* of Harinātha. 14 ff.
656 (5) (5087) *Saṅkrāntiphaladīpikā* of Viśvanātha. 10 ff.
*657 (6) (5130) *Pātasāraṇī* of Gaṇeśa. 6 ff. Saṃ. 1875 (A.D. 1818).

658 (7) (5121) *Dvādaśabhāvaphala.* 2 ff.
659 (8) (4850) *Tājakabhūṣaṇa.* 19 ff.
660 (9) (4888) *Dvādaśabhāvaphala.* 5 ff. Saṃ. 1910, Śaka 1775 (A.D. 1853).
661 (10) (5150) *Yoginīdaśāntardaśāphala.* 2 ff. Copied by Viṣṇudatta in Saṃ. 1892 (A.D. 1835).
662 (11) (4915) *Śivājñā* (?). 4 ff.
663 (12) (5123) *Dvādaśabhāvavicāra.* 46 ff.
664 (13) (5106) *Grahasaṃjñā.* 3 ff. Saṃ. 1849, Śaka 1714 (A.D. 1792).
*665 (14) ( ) *Arghakāṇḍa.* 14 ff.
*666 (15) (4948) Unidentified astrological text. 3 ff.
667 (16) (4829) *Jātakālaṅkāra.* 4 ff.
668 (17) (5206) *Gaurījātaka.* 5 ff. Saṃ. 1846 (A.D. 1789).
669 (18) (5133) *Pārāśarīdaśā.* 7 ff.
670 (19) (5009) *Nīlakaṇṭhī* of Nīlakaṇṭha; this manuscript contains only the *Saṃjñātantra.* 15 ff. Saṃ. 1855, Śaka 1720 (A.D. 1798).
671 (20) (4787) *Yogaśataka* of Śrīdaivajña. 7 ff.
672 (21) (4703) *Jātakālaṅkāra* of Gaṇeśa. 15 ff. Saṃ. 1863 (A.D. 1806).
673 (22) (5281) *Svapnādhyāya.* 6 ff. Saṃ. 1881 (A.D. 1824).
*674 (23) (5131) *Uḍudāyapradīpa* of Parāśara. 8 ff.
675 (24) (5124) *Dvādaśarāśisvarūpana* with a commentary. 2 ff.
676 (25) (4704) *Jātakālaṅkāra* of Gaṇeśa. 10 ff.
677 (26) (4977) *Jātakābharaṇa* of Ḍhuṇḍhirāja and *Ṣaṭsaptavargavicāra.* 25 ff. Copied by Raghunātha in Śaka 1752 (A.D. 1830).
678 (27) (5230) *Jayalakṣmī,* a commentary by Harivaṃśa on Narapati's *Narapatijayacaryā.* 3 ff.
679 (28) (5115) *Camatkāracintāmaṇi.* 13 ff. Saṃ. 1745 (A.D. 1688).
*680 (29) (4908) *Līlāvatī* of Bhāskara II. ff 1–6a and 6b–9.
*681 (30) (5053) *Pramitākṣarā,* a commentary by Rāma on his own *Muhūrtacintāmaṇi.* 29 ff.
*682 (31) ( ) A further part of the preceding manuscript. ff. 33–38.
*683 (32) (5294) *Rasasaṅgrahasiddhānta* of Govindarāma; on materia medica. 79 ff. Saṃ. 1919 (A.D. 1862).
684 (33) (4707) *Jātakālaṅkāra* of Gaṇeśa with a commentary. ff. 14–21 and 29–53.
685 (34) (4705) *Jātakālaṅkāra* of Gaṇeśa. 12 ff. Copied by Govardhana Bhaṭṭa in Saṃ. 1889 (A.D. 1832).
686 (35) (4963) *Śīghrabodha* of Kāśīnātha. ff. 11–23.
687 (36) (5234) *Svapnādhyāya* of Bṛhaspati. 2 ff.
688 (37) (5118) *Jātakābharaṇa.* 9 ff.
689 (38) (4994) *Muhūrtamārtaṇḍa* of Nārāyaṇa. 23 ff.
690 (39) (5000) *Nīlakaṇṭhī* of Nīlakaṇṭha. 23 ff. Śaka 1699 (A.D. 1777).
691 (40) (5042) *Muhūrtacintāmaṇi* of Rāma. 38 ff.
692 (41) (4806) *Aṣṭakavargasṛṣṭi.* 3 ff.
*693 (42) (5050) *Muhūrtacintāmaṇi* of Rāma with a commentary. 7 ff.
*694 (43) (4494) *Ahībalacakra* from the *Brahmayāmala.* 2 ff.
*695 (44) (4809) *Kālacakra.* 13 ff. and 1 f.
696 (45) (5151) *Yoginīdaśāphala.* 9 ff.
*697 (46) (4986) *Pātasāraṇī* of Gaṇeśa with the commentary of Dinakara. 9 ff.
698 (47) (4713) *Praśnatattva* of Cakrapāṇi. 12 ff. Saṃ. 1897, Śaka 1762 (A.D. 1840).
699 (48) (4685) *Grahalāghava* of Gaṇeśa. 17 ff.
700 (49) (4835) *Upadeśasūtra* of Jaimini. 16 ff.
701 (50) (5096 and 4885) *Aṣṭottarīdaśākrama* and *Mahādaśāntardaśāvidaśopadaśāphalāni.* 16 ff.
*702 (51) (5171) *Tithicintāmaṇi* of Gaṇeśa. 16 ff.
703 (52) (5007) *Nīlakaṇṭhī* of Nīlakaṇṭha; this manuscript contains only the *Varṣatantra.* 41 ff. Saṃ. 1876 (A.D. 1819).
704 (53) (4898) *Rājāvalī.* 12 ff. Copied by Bālambhaṭṭa in Saṃ. 1924, Śaka 1789 (A.D. 1867).
*705 (54) (4896) Anonymous of 1520. 12 ff.
706 (55) (4834) *Upadeśasūtra* of Jaimini. 4 ff.
707 (56) (4706) *Jātakālaṅkāra* of Gaṇeśa. 1 f.
708 (57) (4890) *Yantrabhedadarśana.* 3 ff.
*709 (58) (4854) *Tithicintāmaṇi* of Gaṇeśa. 12 ff.
710 (59) (4978) *Jātakābharaṇa* of Ḍhuṇḍhirāja. 140 ff. Copied by Keśa(va)bhaṭṭa.
711 (60) (4674) *Lagnacandrikā* of Kāśīnātha. ff. 2–57.
712 (61) (5005) *Nīlakaṇṭhī* of Nīlakaṇṭha; this manuscript contains only the *Varṣatantra.* ff. 2–57. Saṃ. 1864 (A.D. 1807).
713 (62) (5162) *Śakunavantī.* 4 ff.

All of these texts which can be identified can also be localized in Northern and Western India. The same is true of the texts represented in the next, and longer series; though some of these manuscripts may be even more precisely tracked down.

1766 (1) (5220) *Gaurījātaka.* 25 ff. Śaka 1793 (A.D. 1871).
1767 (2) (5257) *Pallīpatanakārikā.* 5 ff.
*1768 (3) (1575) *Śakunāvalī* from the *Skandapurāṇa.* 8 ff.
*1769 (4) (4671) *Jñānamañjarī* of Ṛṣiśarman. 23 ff. Saṃ. 1925 (A.D. 1868).
1770 (5) (4964) *Śīghrabodha* of Kāśīnātha. 47 ff. Saṃ. 1882 (A.D. 1825).
1771 (6) (5158) *Lomaśasaṃhitā.* 16 ff.
1772 (7) (5201) *Strījātakas* from the *Horāsāra* (ch. 25), the *Gargajātaka,* Kalyāṇavarman's *Sārāvalī* (ch. 45), and Mīnarāja's *Vṛddhayavanajātaka* (ch. 58–62). 15 ff.
1773 (8) (5258) *Pallīpatanakārikā.* 3 ff.

1774 (9) (5265) *Pallīśaraṭavicāra*. 2 ff.
1775 (10) (5134) *Pārāśarīhorāpaddhati*. 4 ff. Saṃ. 1849 (A.D. 1792).
1776 (11) (5157) *Lomaśasaṃhitā*. 9 ff.
1777 (12) (5196) *Keśavīpaddhatyudāharaṇa*, a commentary by Viśvanātha on Keśava's *Keśavīpaddhati*. 39 ff. Saṃ. 1811 (A.D. 1754).
1778 (13) (4710) *Jñānamañjarī* of Gadādhara (Ṛṣiśarman?). 24 ff.
*1779 (14) (5167) *Bhāvādhyāya* of Bādarāyaṇa. 8 ff. Copied by Śrīdhara Josī.
*1780 (15) (5031) *Praśnavidyā* of Bādarāyaṇa with the commentary of Utpala. 8 ff.
1781 (16) (5175) *Spaṣṭajanmakālabīja*. 2 ff.
*1782 (17) (4516) *Praśnabhairava* of Veṅkaṭeśa. 7 ff. Saṃ. 1923, Śaka 1788 (A.D. 1866).
1783 (18) (4899) *Rājāvalī* (?). 5 ff. Saṃ 1908 (A.D. 1851).
1784 (19) (4900) *Rājāvalī*. 11 ff.
1785 (20) (4686) *Grahalāghava* of Gaṇeśa. 7 ff.
1786 (21) (5062) *Bṛhajjātaka* of Varāhamihira. 23 ff. Saṃ. 1834 (A.D. 1777).
1787 (22) (4997) *Muhūrtamārtaṇḍa* of Nārāyaṇa with a commentary. 52 ff.
*1788 (23) (4882) *Makaranda* of Makaranda. 38 ff.
1789 (24) (4999) *Mārtaṇḍavallabhā*, a commentary by Nārāyaṇa on his own *Muhūrtamārtaṇḍa*. 3 ff.
*1790 (25) (4598) *Akṣaraikāvalī* from the *Rudrayāmala*. 5 ff.
1791 (26) (4687) *Grahalāghava* of Gaṇeśa. 24 ff. Copied by Śivarāma in Śaka 1781 (A.D. 1859).
1792 (27) (5136) *Praśnacūḍāmaṇi*. 23 ff.
1793 (28) (5063) *Bṛhajjātaka* of Varāhamihira. 5 ff.
1794 (29) (5043) *Muhūrtacintāmaṇi* of Rāma. 63 ff. Saṃ. 1925, Śaka 1790 (A.D. 1868).
1795 (30) (4757) *Līlāvatī* of Bhāskara II. 6 ff.
1796 (31) (5145) *Bālabodhinī*. 5 ff. Saṃ. 1930 (A.D. 1873), Śaka 1750 (A.D. 1828) (read Śaka 1795).
1797 (32) (4893) *Yogasaṃgraha*. 12 ff.
1798 (33) (4688) *Grahalāghava* of Gaṇeśa. 10 ff.
*1799 (34) (4856) *Tithicintāmaṇi* of Gaṇeśa. 27 ff.
1800 (35) (4726) *Pañcāṅgasiddhi* of Bābā. ff. 3–7. Copied by Vijñāgadādhara, the son of Lakṣmīdhara, in Śaka 1571 (A.D. 1649).
1801 (36) (5208) *Gaurījātaka*. 38 ff.
*1802 (37) (4796) *Gaṇitanāmamālā* of Haridatta. 6 ff. Copied by Śāmabhaṭṭa Phaḍake in Saṃ. 1839, Śaka 1704 (A.D. 1782).
1803 (38) (5061) *Jaganmohana* of Lakṣmaṇa. ff. 52–278.
1804 (39) (4965) *Śīghrabodha* of Kāśīnātha. 38 ff. Saṃ. 1823 (A.D. 1766).
1805 (40) (5197) *Keśavīpaddhatyudāharaṇa*, a commentary by Viśvanātha on Keśava's *Kesavīpaddhati*. 2 ff.
1806 (41) (5013) *Pārāśara* of Parāśara. 4 ff.
1807 (42) (4961 and 5155) *Lagnacandrikā* of Kāśīnātha and *Laghucāṇakyajātaka* (?). 56 ff. Saṃ. 1832 (A.D. 1775).
1808 (43) (5000a) *Nīlakaṇṭhī* of Nīlakaṇṭha. 42 ff.
*1809 (44) (5084) *Hāyanaratna* of Balabhadra. 19 ff.
*1810 (45) (5085) *Hāyanaratna* of Balabhadra. 18 ff.
1811 (46) (5272) *Ramalasāra*. 10 ff.
1812 (47) (4758) *Līlāvatī* of Bhāskara II. ff. 7–40.
1813 (48) (4746) *Golādhyāya* from the *Siddhāntaśiromaṇi* of Bhāskara II. 115 ff. Saṃ. 1892 (A.D. 1835).
1814 (49) (5027) *Ṣaṭpañcāśikāvivṛti*, a commentary by Utpala on Pṛthuyaśas' *Saṭpañcāśikā*. 5 ff.
1815 (50) (5141) *Praśnabhairava*. 19 ff.
1816 (51) (5231) *Mayūracitraka* of Nārada. 25 ff.
1817 (52) (4930) *Sūryasiddhānta* with a commentary. 11 ff.
1818 (53) (5044) *Muhūrtacintāmaṇi* of Rāma. 10 ff.
1819 (54) (4955) *Muhūrtamārtaṇḍa* of Nārāyaṇa. 24 ff. Śaka 1766 (A.D. 1844).
*1820 (55) (4818) *Grahalalita*. 10 ff. Śaka 1776 (A.D. 1854).
1821 (56) (5156) *Laghujātaka*. 19 ff.
1822 (57) (4689) *Grahalāghava* of Gaṇeśa. 34 ff. Śaka 1738 (A.D. 1816).
1823 (58) (5279) *Sāmudrika*. 11 ff.
*1824 (59) (5108) Anonymous of 1520. ff. 2–10.
1825 (60) (5010) *Nīlakaṇṭhī* of Nīlakaṇṭha. This manuscript contains only the *Saṃjñātantra*. 16 ff.
1826 (61) (5242) *Samarasāraṭīkā*, a commentary by Bharata on Rāmacandra's *Samarasāra*. 32 ff. Copied by Sadāśiva in Śaka 1681 (A.D. 1759).
*1827 (62) (5132) *Uttarabhāga* of the *Parāśarahorā* of Parāśara. 37 ff.
1828 (63) (4966) *Śīghrabodha* of Kāśīnātha. 34 ff. Copied by Rājarāma, the son of Gaṅgādhara.
1829 (64) (5038) *Bālabodha* of Muñjāditya. 14 ff.
1830 (65) (5039) *Bālabodha* of Muñjāditya. 22 ff.
1831 (66) (5226) *Narapatijayacaryā* of Narapati. 73 ff.
*1832 (67) (4953 and 5203) *Santānadīpikā*, *Praśnavaiṣṇava* of Nārāyaṇadāsa Siddha. and *Kālajātaka*. 6 ff. Saṃ. 1899, Śaka 1766 (read 1764) (A.D. 1842).
1833 (68) (5250) *Jayalakṣmī*, a commentary by Harivaṃśa on Narapati's *Narapatijayacaryā*. 34 ff.
1834 (69) (4933) *Sūryasiddhāntodāharaṇa*, a commentary by Viśvanātha on the *Sūryasiddhānta*. 212 ff.
1835 (70) (4714) *Praśnavidyā* of Caṇḍeśvara. ff. 21–104. Saṃ. 1857 (A.D. 1800).
1836 (71) (4847) *Jyotiḥsāgara*. ff. 9–61. Copied by Rāmacandra Cāpekara in Śaka 1712 (A.D. 1790).
*1837 (72) (4881) *Makaranda* of Makaranda. 20 ff.
*1838 (73) (4880) *Makarandavivaraṇa*, a commen-

tary by Divākara on Makaranda's *Makaranda.* 2 ff.

1839 (74) (4831) *Jātakālaṅkāra.* 6 ff.

1840 (75) (4830) *Jātakālaṅkāra.* 17 ff. Copied by Śivarāma in Saṃ. 1926, Śaka 1791 (A.D. 1869).

1841 (76) (5014) *Pārāśarī.* ff. 1–5 and 3 ff.

1842 (77) (4989) *Karmaprakāśikāvṛtti,* a commentary by Nārāyaṇa Sāmudrika on Samarasiṃha's *Manuṣyajātaka.* 10 ff.

*1843 (78) (5211) *Grahalāghavodāharaṇa,* a commentary on Gaṇeśa's *Grahalāghava.* 5 ff.

1844 (79) (4990) *Karmaprakāśikāvṛtti,* a commentary by Nārāyaṇa Sāmudrika on Samarasiṃha's *Manuṣyajātaka.* 24 ff.

*1845 (80) (5082) *Laghujātaka* of Varāhamihira with the commentary of Utpala. 46 ff. Saṃ. 1833 (A.D. 1776).

1846 (81) (4725) *Kālāmṛtavivṛti* of Bāpirāja. 11 ff.

*1847 (82) (4905) *Tithicintāmaṇi* of Gaṇeśa. 23 ff.

*1848 (83) (4907) *Tithicintāmaṇi* of Gaṇeśa. 23 ff.

*1849 (84) (5182) Unidentified astrological text. 7 ff. Śaka 1649 (A.D. 1727).

1850 (85) (5138) *Praśnatantra.* 13 ff. Śaka 1673 (A.D. 1751).

*1851 (86) (4947) Unidentified astrological text. 4 ff.

1852 (87) (5164) *Śakunasaṃgraha.* 4 ff.

1853 (88) (5012b) *Nīlakaṇṭhyudāharaṇa* (?), a commentary by Viśvanātha on Nīlakaṇṭha's *Nīlakaṇṭhī.* 17 ff.

1854 (89) (5139) *Praśnatantra* with a commentary. 6 ff.

1855 (90) (4852 and 5128) *Tārābalābala.* 6 ff. Saṃ. 1867 (A.D. 1810) and *Padmakośa.* 12 ff. Saṃ. 1889 (A.D. 1832).

*1856 (91) (5030) *Hāyanaratna* of Balabhadra. 11 ff.

1857 (92) (5066) *Bṛhajjātaka* of Varāhamihira with the commentary of Utpala. 18 ff.

1858 (93) (5045) *Muhūrtacintāmaṇi* of Rāma. 40 ff. The data Śaka 1522 (A.D. 1600) is that of the composition of the work, not of the copying of the manuscript.

*1859 (94) (4814) *Tithicintāmaṇi* of Gaṇeśa. 12 ff. Copied by Cīreśvara Mahājana in Saṃ. 1854 (A.D. 1797).

1860 (95) (5240) *Samarasāra* of Rāmacandra. 7 ff. Śaka 1672 (A.D. 1750).

1861 (96) (5105) *Kautukalīlāvatī.* ff. 1–2 and 4.

1862 (97) (5227) *Narapatijayacaryā* of Narapati. 2 ff.

1863 (98) (5209) *Grahagocara.* 3 ff.

1864 (99) (5094) *Saṅketakaumudī* of Harinātha. 7 ff.

1865 (100) (5273) Unidentified astrological text. 4 ff.

1866 (101) (4786) *Saṃkrāntiprakaraṇa* of Śiva. 19 ff.

1867 (102) (5000b) *Nīlakaṇṭhī* of Nīlakaṇṭha. 50 ff.

*1868 (103) (4919) Unidentified astrological text. 16 ff.

1869 (104) (4676) *Jyotiṣkedāra* of Kṛpāśaṅkara. 46 ff.

1870 (105) (5046) *Muhūrtacintāmaṇi* of Rāma. 43 ff.

1871 (106) (5235 and 4959) *Svapnādhyāya* of Bṛhaspati and *Navagrahastotra* of Kālidāsa. 5 ff.

1872 (107) (5195) *Keśavīpaddhati* of Keśava. 47 ff. Śaka 1720 (A.D. 1798).

1873 (108) (5169) *Saṃvatsaraphala.* ff. 22–23. Śaka 1728 (A.D. 1806).

1874 (109) (4901) *Rājāvalī.* 21 ff.

1875 (110) (5270) *Ramalavidyā.* 6 ff.

1876 (111) (4996) *Muhūrtamārtaṇḍa* of Nārāyaṇa. 28 ff. Saṃ. 1776, Śaka 1641 (A.D. 1719).

*1877 (112) (5172) Anonymous of 1520. 2 ff.

1878 (113) (5210) *Grahagocara.* 6 ff. Saṃ. 1887 (A.D. 1830) or possibly Saṃ. 1807 (A.D. 1750).

1879 (114) (4973) *Pāśakakeralī* of Garga. ff. 6–11.

1880 (115) (5274) *Saṃgrahapraśnavicāra.* 12 ff.

1881 (116) (5135) *Praśnavidyā* of Caṇḍeśvara. 4 ff. Saṃ. 1812 (A.D. 1755).

1882 (117) (4957) *Akṣaracintāmaṇi* of Śiva. 11 ff.

1883 (118) (5137, 5152, 5144, and 5168) *Praśnajñānaviveka, Darśanapraśna, Prasannapraśna,* and *Saṃyogākṣarayukti.* 2 ff.

1884 (119) (5143) *Praśnamāṇikya.* 4 ff.

1885 (120) (? and 5149) ? and *Muhūrtarāja.* ff. 7–9. Saṃ. 1867 (A.D. 1810).

1886 (121) (4793) *Jātakālaṅkārakarman* of Śrīśuka. 4 ff.

1887 (122) (5100) *Āyurdāyavibhāga.* 7 ff.

*1888 (123) (4816) *Grahaṇa* of Mallāri and text on ramala. 4 ff.

*1889 (124) (4910) Text on muthahā (Arabic منتهى) and a *Grahabhāvaphalam.* 11 ff. Copied by Rāmacandra Śukla in Avaṃtigrāma (Ujjain) in Saṃ. 1786 (A.D. 1729).

*1890 (125) (4808) *Bhadrāvicāra* (in upajāti meter) with a commentary. 2 ff.

*1891 (126) (4906) *Tithicintāmaṇi* of Gaṇeśa. 15 ff. Śaka 1683 (A.D. 1761).

1892 (127) (5259) *Pallīpatanaphala.* 2 ff. Copied by Mairāla Bhaṭṭa.

1893 (128) (5104) *Keralījātaka.* 15 ff. Śaka 1797 (A.D. 1875).

1894 (129) (5090) *Muhūrtamuktāvalī* of Paramahaṃsa Parivrājaka. 8 ff. Saṃ. 1836 (A.D. 1779).

*1895 (130) (4860) Geographical Table. 2 ff.

1896 (131) (4795) *Kalpalatā* of Soma. ff. 17–28. Copied by Veṅkaṭa Pāṭhaka.

*1897 (132) (4804) *Ābhrachāyā* with a commentary. 2 ff.

*1898 (133) (4815) Table of sahama (Arabic سهم). 2 ff.

1899 (134) (5285) Unidentified astrological text. 1 f.

*1900 (135) (5238) *Vṛddhayavanajātaka* of Mīna-

rāja; this manuscript contains only chapters 7 and 8. 8 ff. Identical with Oudh (1888) 14.
1901 (136) (5041) *Muhūrtasarvasva* of Raghuvīra. 17 ff. Identical with N-W Prov. IX (1885) 17 (from Benares).
*1902 (137) (5183) *Makarandavivaraṇa,* a commentary by Divākara on Makaranda's *Makaranda.* 3 ff.
1903 (138) (4976) *Praśnamanoramā* of Garga with a commentary. 6 ff. Saṃ. 1832 (A.D. 1775).
1904 (139) (5254) Unidentified text on palmistry. 1 f.
1905 (140) (5255) Unidentified text on palmistry. 1 f.
*1906 (141) (4844) *Jyotiṣasaṃgraha.* 28 ff. Saṃ. 1835 (A.D. 1778).
1907 (142) (5140) *Praśnatilaka.* 3 ff.
*1908 (143) (5161) *Vṛddhayavanajātaka* of Mīnarāja; this manuscript contains chapters 17 to 31. 38 ff.
*1909 (144) (4873) *Pāṭīgaṇita.* ff. 3–8.
1910 (145) (5207) *Gaurījātaka.* 6 ff.
1911 (146) (5122) *Dvādaśabhāvavicāra.* 2 ff. Copied by Moreśvara.
*1912 (147) (4810) *Iṣṭakālajñāna, Lomaśasaṃhitā,* unidentified text on interrogations with a commentary, and chapter 8 of Kalyāṇavarman's *Sārāvalī.* 4 ff. Saṃ. 1897 (A.D. 1840).
*1913 (148) (4799, 4798, and 4955). Anonymous of 1520. 4 ff. Śaka 1798 (A.D. 1876).
*1914 (149) (5969) *Śakunapraśna;* in Hindi. 8 ff.
1915 (150) (5193) *Keśavīpaddhati* of Keśava. 11 ff.
1916 (151) (5000c) *Nīlakaṇṭhī* of Nīlakaṇṭha. 5 ff.

Except for one or two texts—e.g., the *Kālāmṛtavivṛti* of Bāpirāja (1846)—all of these works are commonly met with in Western and Northern India. One manuscript (1848) has the ex libris, he pustaka dāmodara śāstri sahasrabuddhe, penciled on f. 1r; this indicates a Mahārāṣṭrian provenance. And two manuscripts (1900 and 1901), can be traced to Uttar Pradesh; presumably this will be possible with others also. We do not have, in any of our three major collections, any substantial representation of the astronomy and astrology of Southern or of Eastern India; and we are quite secure in assigning all of the tables discussed in this catalogue to Northern and Western India, and particularly to Gujarat, Rajasthan, and Uttar Pradesh.

Part of the supporting evidence for this statement lies in the location of the manuscripts containing copies of the works represented here. I have given lists of all such manuscripts at present known to me; further information, including full bibliographical references to manuscript catalogues and to editions of texts, will be found in my *Catalogue of the Exact Sciences in Sanskrit,* now in the course of preparation. For the moment I hope the reader will be content with the locations of the libraries whose code-names are employed in these lists.

ABSP. Akhila Bharatiya Sanskrit Parishad, Lucknow.
Adyar. Adyar Library, Madras.
Alwar. Palace Library, Alwar, Rajasthan.
Ānandāśrama. Ānandāśrama, Poona.
Anup. Palace Library, Bikaner, Rajasthan.
Baroda. Oriental Institute, Baroda.
Bharatpur. State Library, Bharatpur, Rajasthan.
BORI. Bhandarkar Oriental Research Institute, Poona.
DC. Deccan College, Poona.
Goṇḍal. Bhuvaneśvarī Pīṭh, Goṇḍal, Saurāṣṭra.
Gujarāt Vidyā Sabhā. Gujarāt Vidyā Sabhā, Ahmadabad.
Jaipur. Palace Library, Jaipur, Rajasthan.
Kotah. State Library, Kotah, Rajasthan.
LDI. LD Institute, Ahmadabad.
RORI. Rajasthan Oriental Research Institute, Jodhpur, Rajasthan.
SOI. Scindia Oriental Institute, Ujjain.
Tod. Royal Asiatic Society, London (collected in Rajasthan).
Vārāṇasī. Saravatī Bhavana, Benares.
VVRI. Vishveshvaranand Vedic Research Institute, Hoshiarpur, Panjab.

CATALOGUE

4675 (Smith Indic 185 ff. 42–55). See 4877.

4678 (Smith Indic 94 (a) ff. 4v–15v). See 4679.

4679, 4678, 4681, 4682 (Smith Indic 94) 40 ff. This manuscript is in three sections: (a) ff. 1–16, (b) ff. 1–12, and (c) ff. 1–12. On (a) f. 1r is written: naṃ 19 pattra 55, and, in a different hand: N 27. On (a) f. 4r is written: śake 1652 sādhāra śrāvaṇa śu 7. This date corresponds to 10 July 1730. On (a) f. 1r, at the top, is written: ślo 310. On (c) f. 12v, at the bottom right-hand corner, is written, in European numerals: 6537. (a) f. 1 is supplied by a later hand.

(a) f. 1r (4679). Laghucintāmaṇi.
Ff. 1v–4r. Text of the *Tithicintāmaṇi.*
Ff. 4v–15v (4678). Table 1 of the *Tithicintāmaṇi* for 0 to 388 tithis.
Ff. 16r–16v. Blank.

(b) ff. 1r–12r (4681). Table 2 of the *Tithicintāmaṇi* for 0 to 390 nakṣatras.
F. 12v. Blank.

(c) Ff. 1r–12v (4682). Table 3 of the *Tithicintāmaṇi* for 0 to 408 yogas.

4680 (Smith Indic 92) 12 ff. On f. 4v is written: śake 1704 śubhakṛtasaṃvatsare adhikajeṣṭa śuddha 13 trayodaśī śanivāsare saṃpurṇa. This date corresponds to Saturday, 25 May 1782. On f. 12r is written: śake 1704 śubhakṛtanāmasaṃvatsare jeṣṭa śuddha tritiyā guruvāsare. This date corresponds to Thursday, 13 June 1782. On f. 1r, at the top, is written: śloka 190, and, at the bottom, in European numerals: 6607.
F. 1r. Laghucintāmaṇi prāraṃbhaḥ.
Ff. 1v–3v. Text of the *Tithicintāmaṇi* (20 verses).
Ff. 3v–4v. Brief comments on the *Tithicintāmaṇi*.
Ff. 5r–12r. Text of the *Bṛhattithicintāmaṇi*.
F. 12v. Laghucintāmaṇi saṃpūrṇa.

4681 (Smith Indic 94 (b) ff. 1–12). See 4679.

4682 (Smith Indic 94 (c) ff. 1–12). See 4679.

4694 (Smith Indic 26) 6 ff. On f. 1r, in the upper left, is written: ⟨śloka⟩ 150; on f. 6v, in the left-hand margin, is written: śloka 200.
Ff. 1r–3r. Set A of the *Grahalāghavasāriṇī*.
Ff. 3r–3v. Set B of the *Grahalāghavasāriṇī*.
Ff. 3v–4r. Set C of the *Grahalāghavasāriṇī*.
Ff. 4r–4v. Set D of the *Grahalāghavasāriṇī*.
F. 5r. Set E of the *Grahalāghavasāriṇī*.
F. 5v. Table of planetary aspects.
F. 5v. Table of the extents (in minutes) of horās (900), drekāṇas (600), saptāṃśas (257), navamāṃśas (200), and dvādaśāṃśas (150).
F. 5v. Table of planetary fines.
F. 6r. Table of planetary friends.
F. 6r. Astrological table.
F. 6r. Table of the dejections of the planets.
F. 6v. Table of the longitudes of the nakṣatras and of the planets associated with them.
F. 6v. Table of aṣṭottarīdaśās and of yoginīs.

4708 (Smith Indic 16) 39 ff. On f. 1r is written: tithicintāmaṇikoṣṭhakāni, and: ślo 800.
Ff. 1v–39v. Table combining tables 1, 2, and 3 of the *Bṛhattithicintāmaṇi* for 1 to 4 cycles and for k = 0 to k = 138 of the 5th cycle.

4709 (Smith Indic 151) 107 ff. This manuscript is in 4 sections: (a) 2 ff., (b) 40 ff., (c) 32 ff., and (d) 33 ff. On (d) f. 33r, at the bottom, is written: śake 1682 vikramanāmasaṃvatsare phālguna vadi 8 śanau taddine idaṃ pustakaṃ nārāyaṇena likhitaṃ svārthaṃ paropakārārthaṃ ca śubham astu. This date corresponds to Saturday, 14 March 1761. On (a) f. 1r is written: ślo 5000.

(a) ff. 1v–2v. First 25 verses of the *Bṛhattithicintāmaṇi*.

(b) ff. 1r–40r. Table 1 of the *Bṛhattithicintāmaṇi* for 27 cycles.
F. 40v. Blank.

(c) ff. 1r–32v. Table 2 of the *Bṛhattithicintāmaṇi* for 27 cycles.

(d) ff. 1r–33r. Table 3 of the *Bṛhattithicintāmaṇi* for 27 cycles.
F. 33v. Blank.

4716 (Harvard 525) 5 ff.
F. 1r. Tables of the positions of the Lord of the Year, the Epact, the Moon, and the Lunar Anomaly for Śaka 1634, 1635, 1636 (Lunar Node included), 1637, 1621, and 1618 (A.D. 1712, 1713, 1714, 1715, 1699, and 1696).
Ff. 1v–5r. Text of the *Candrārkī* of Dinakara.
Ff. 5r–5v. Scholia on the *Candrārkī's* parameters.
F. 5v. Table of parameters and epoch positions of the Lord of the Year, the Epact, and the Tithikendra; *cf.* table 5 of the *Candrārkī*.

	Yearly motion	*Epoch position*	*Bīja*
Lord of the Year	1;15,31,17,17 days	1;24,57,12,30 days	
Epact	11;3,53,22,40 tithis	22;19,34,10,0 tithis	−0;1,15
Tithikendra	7;40,30,44,0 tithis	21;52,16,55,0 tithis	−0;3,45

4717, 4923, 4824, 4823 (Smith Indic 190) ff. 3–6 and 8–17.
Ff. 3r–3v (4717) contain verses 20–26 (the concluding verses) of the text of the *Candrārkī*. On f. 3v after verse 26: iti śrīdinakarācāryyaviracitā candrārkī saṃpūrṇā.
Ff. 3v–6v (4923). Table 1 of the *Candrārkī* for 0 to 365 days.
Ff. 8r–12v (4824). Table 2 of the *Candrārkī* for 62 to 365 days.
Ff. 13r–17v (4823). Table 4 of the *Candrārkī* for 0° to 299°.

4718 (Smith Indic 53) 2 ff. On f. 1r, on the upper left-hand corner, is written: ślo 50.
Ff. 1r–1v. Text of the *Tithyādicintāmaṇi* (12 verses).
F. 1v. Table 1 of the *Tithyādicintāmaṇi*.
F. 2r. Table 2 of the *Tithyādicintāmaṇi*.
F. 2r. Table 3 of the *Tithyādicintāmaṇi*.
F. 2v. Blank.

4720 (Smith Indic 49) 19 ff. On f. 18r is written: liḥ gaphuramaṇi tripāṭika svāsvārthaṃ parārthaṃ vā kāśyāṃ govardhane pure saṃ 1922 āśvina kṛṣṇa tithau 14. This date corresponds to 17 October 1865.
F. 1r. Blank.
Ff. 1v–18r. The *Makarandavivaraṇa* of Divākara. The colophon (on f. 18r) is: iti śrīsakalagaṇakasārvabhaumasuṭānandadaivajñasutanṛsiṃhadaivajñasūnunā divākareṇa viraci⟨taṃ⟩ makarandavivaraṇaṃ samāptam.
F. 17v. Table of aṣṭottarī and viṃśottarī rājās; similar to table 5 of the Anonymous of 1594.
Ff. 18v–19v. Blank.

4721 (Smith Indic 79) ff. 6–7. On f. 7v is written:

saṃvat 1853 śāke 1718 māsānāṃ māsottame (?) māse kṛṣṇe pakṣe kṛṣṇasya ekādasyāṃ śanivāsare tadine liṣitaṃ pūstakaṃ svārthaṃ parārthaṃ vā iti. This date corresponds to Saturday, ? 1797. On f. 7v, in the lower right-hand corner, is written, in European numerals: 6526.

Ff. 6r–7v. The last 23½ verses of the *Makarandavivaraṇa* of Divākara. The colophon reads: iti śrīsakalagaṇakasarvabhaumaśrīkṛṣṇadaivajñasūtanṛsiṃhasya sutena divākareṇa viracitaṃ makarandavivaraṇaṃ samāptam.

4735 (Smith Indic 45) ff. 2–10 and 12–17. On f. 17v, in the left-hand margin, is written: iti śrībrahmatulyasya grahasādhanārthaṃ sāranī sa⟨māptā⟩; in the right-hand margin: sāraṇī samāpta // saṃvat 1855 varṣe śāke 1720 pravarttamani kārttika vida 11 some. This data corresponds to Monday, 3 December 1798. On f. 2r, at the top, is written: śloka 600.

F. 2r. Table 1 of the *Brahmatulyasāraṇī*. In this manuscript the longitudes for 1 to 30 periods of 20 "years" are omitted.
F. 2v. Table 2 of the *Brahmatulyasāraṇī*.
F. 3r. Table 3 of the *Brahmatulyasāraṇī*.
F. 3v. Table 4 of the *Brahmatulyasāraṇī*.
F. 4r. Table 5 of the *Brahmatulyasāraṇī*.
F. 4v. Table 6 of the *Brahmatulyasāraṇī*.
F. 5r. Table 7 of the *Brahmatulyasāraṇī*.
F. 5v. Table 8 of the *Brahmatulyasāraṇī*.
F. 6r. Table 9 of the *Brahmatulyasāraṇī*.
F. 6v. Table 10 of the *Brahmatulyasāraṇī*.
F. 7r. Table 11 of the *Brahmatulyasāraṇī*.
F. 7v. Table 12 of the *Brahmatulyasāraṇī*.
F. 7v. Table 13 of the *Brahmatulyasāraṇī*.
F. 8r. Table 14 of the *Brahmatulyasāraṇī*.
Ff. 8v–9v. Table 15 of the *Brahmatulyasāraṇī*.
F. 10r. Table 16 of the *Brahmatulyasāraṇī*.
Ff. 10v–11v (ff. 11r–11v = 4946 LXII). Table 17 of the *Brahmatulyasāraṇī*.
F. 12r. Table 18 of the *Brahmatulyasāraṇī*.
Ff. 12v–13v. Table 19 of the *Brahmatulyasāraṇī*.
F. 14r. Table 20 of the *Brahmatulyasāraṇī*.
Ff. 14v–15v. Table 21 of the *Brahmatulyasāraṇī*.
F. 16r. Table 22 of the *Brahmatulyasāraṇī*.
Ff. 16v–17v. Table 23 of the *Brahmatulyasāraṇī*.

4763 (Smith Indic 192) 29 ff. On f. 1r is written, in the upper margin: śloka 1500; on f. 15v, in the lower margin: iti śrīmahādevoktaṃ baumapaṃkti 59 saṃpūrṇā.

Ff. 1r–15v. Table 9 of the *Mahādevī* (N = 0 to 59); columns A, B, D, and F, and function G.
Ff. 16r–29v. Table 10 of the *Mahādevī* (N = 0 to 55); columns A, B, D, and F, and function G.

4763a (Smith Indic 41) 128 ff. (separate numeration for each planet). On f. 30v of Saturn, at the bottom, is written: śrīgopī janavallabho vijayate // // śāke 1588 pravarttamāne āṣāḍhamāse śuddhapakṣe tṛtīyāyāṃ śanidine likhiteyaṃ mahādevī viṭṭhalajinā //. This date corresponds to Saturday, 23 June 1666 in the Julian calendar. In the upper margin of f. 30v is written: śloka 3000.

Mars. Ff. 9–18, 28, and 30. Table 9 of the *Mahādevī* (N = 16 to 35,54,55,58, and 59).
Mercury. Ff. 4 and 6–30. Table 10 of the *Mahādevī* (N = 6,7, and 10 to 59).
Jupiter. Ff. 1–30. Table 11 of the *Mahādevī* (N = 0 to 59).
Venus. Ff. 1–30. Table 12 of the *Mahādevī* (N = 0 to 59).
Saturn. Ff. 1–30. Table 13 of the *Mahādevī* (N = 0 to 59).

4763a (Smith Indic 41) additional folio.
Mercury. F. 31 is from a different manuscript.
F. 31r. Table 2 of the Anonymous III of 1578.
Ff. 31r–31v. Commentary on the *Mahādevī* describing the use of function G and of column F.

4764 (Smith Indic 80) 9 ff. On f. 9v, in the upper margin, is written: śloka 200; in the lower margin, in European numerals: 6570. All tables are specified as coming from the *Mahādevī* save those on ff. 1v, 3r, and 3v.

Ff. 1r–1v. Table 2 of the *Mahādevī*. Facsimile in *Proc. Amer. Philos. Soc.* 111, 1967, 70.
Ff. 2r–2v. Table 1 of the *Mahādevī*.
Ff. 3r–3v. Table 2a (of the *Mahādevī*?).
Ff. 4r–4v. Table 3 of the *Mahādevī*.
Ff. 5r–5v. Table 4 of the *Mahādevī*.
Ff. 6r–6v. Table 5 of the *Mahādevī*.
Ff. 7r–7v. Table 6 of the *Mahādevī*.
Ff. 8r–8v. Table 7 of the *Mahādevī*.
Ff. 9r–9v. Table 8 of the *Mahādevī*.

4765 (Smith Indic 22) 3 ff. On f. 1r, in the upper margin, is written: ⟨ślo⟩ 120; on f. 3v, in the upper margin: śloka 120.

F. 1r. Table 3 of the *Mahādevī*.
F. 1v. Table 4 of the *Mahādevī*.
F. 2r. Table 5 of the *Mahādevī*.
F. 2v. Table 6 of the *Mahādevī*.
F. 3r. Table 7 of the *Mahādevī*.
F. 3v. Table 8 of the *Mahādevī*.

4766 (Smith Indic 139) 138 ff. On f. 1r, in the upper margin, is written: śloka 4500. On f. 65v, in the lower margin, is written: . . . trivikrahmādvijainaivaḥ gaṇikānāṃ sukhāptaye; on f. 122v, in the upper margin: . . . trivikramadvijenaiva gaṇakānāṃ sukhāptaye.

Ff. 1r–20v. Table 8 of the Anonymous of 1704 (N = 0 to 78). The following years are noted: N = 0 (f. 1r), 1626 (A.D. 1704); N = 4 (f. 2r), 1709 (A.D. 1787).
Ff. 21ar–21av and 21br–44v. Table 9 of the Anonymous of 1704 (N = 0 to 45).
Ff. 45r–65v. Table 10 of the Anonymous of 1704 (N = 0 to 82). The following years are noted: N = 0 (f. 45r), 1709 (A.D. 1787); N = 74 (f. 63v), ⟨1⟩700 (A.D. 1778);

N = 76 (f. 64r), 1702 (A.D. 1780); N = 82 (f. 65v), śā⟨ka⟩ 1708 (A.D. 1786).
Ff. 66r–122v. Table 11 of the Anonymous of 1704 (N = 0 to 226).
Ff. 123r–137v. Table 12 of the Anonymous of 1704 (N = 0 to 58). The following years are noted: N = 0 (f. 123r), 1744 (A.D. 1822); N = 2 (f. 123v), 1746 (A.D. 1824); N = 5 (f. 124r), 1690 (A.D. 1768); N = 6 (f. 124v), 1691 (A.D. 1769); N = 7 (f. 124v), 1692 (A.D. 1770); N = 8 (f. 125r), 1693 (A.D. 1771); N = 11 (f. 125v), 1696 (A.D. 1774); N = 15 (f. 126v), 1700 (A.D. 1778); N = 16 (f. 127r), 1701 (A.D. 1779); N = 19 (f. 127v), 1704 (A.D. 1782); N = 22 (f. 128v), 1707 (A.D. 1785); N = 24 (f. 129r), 1709 (A.D. 1787); N = 28 (f. 130r), 1713 (A.D. 1791); N = 29 (f. 130r), 1714 (A.D. 1792); N = 30 (f. 130v), 1715 (A.D. 1793); N = 39 (f. 133r), 1724 (A.D. 1802); N = 46 (f. 134v), 1731 (A.D. 1809); N = 47 (f. 135r), 1732 (A.D. 1810); N = 49 (f. 135v), 1734 (A.D. 1812); N = 55 (f. 137r), 1740 (A.D. 1818); N = 57 (f. 137v), 1742 (A.D. 1820).

4767 (Smith Indic 18) ff. 1 and 3–7. On f. 1r, at the bottom and in a different hand, is written: mahādevīkoṣṭakāṇi. On f. 7v, in the upper margin, is written: 180 śloka; in the lower margin: iti dhārkādāsanāṃ pānābe laṣit galālavije.
Ff. 1r–1v. Table 1 of the *Mahādevī*.
Ff. 3r–3v. Table 3 of the *Mahādevī*.
Ff. 4r–4v. Table 4 of the *Mahādevī*.
Ff. 5r–5v. Table 5 of the *Mahādevī*.
Ff. 6r–6v. Table 6 of the *Mahādevī*.
Ff. 7r–7v. Table 7 of the *Mahādevī*.

4768 (Harvard 218) 5 ff. On f. 5v is written: saṃvat 1873 vaiśākhaśuklā pratipadyāṃ śanau / śāke 1738. This date corresponds to Saturday, 27 April 1816.
F. 1r. Blank.
Ff. 1v–5v. The *Grahasiddhiṭīkā*, a commentary on Mahādeva.

4769, 4773 (Smith Indic 147) 60 ff. plus 1 unnumbered f. On f. 1r is written: mahādevī. On the verso of the unnumbered f. is written: śloka 1100.
Ff. 1r–30v (4769). Table 9 of the *Mahādevī* (N = 0 to 59).
Ff. 31r–60v (4773). Table 10 of the *Mahādevī* (N = 0 to 59).
Unnumbered f., recto. Table 10 of the *Mahādevī* (N = 30).
Unnumbered f., verso. Blank.

4770 (Smith Indic 179) ff. 3–16. On f. 3r, in the upper right-hand corner, is written: ślo 250, in the lower right-hand corner in European numerals: 6575.
F. 3r. Verses 36–43, which conclude the *Grahasiddhi* of Mahādeva.
F. 3r. Text on using the cālakas of the planets.
F. 3v. Table of bījas to the epoch positions of the *Mahādevī*.
F. 3v. Table of yearly parameters and of epoch positions of the *Mahādevī* for Epact, Lord of the Year, Mars, Mercury, Jupiter, Venus, Saturn, and the Lunar Node.
F. 3v. Table of motion of the Lunar Node for 1 to 27 (= 0 to 26) avadhis. The daily motion is precisely −0;3,11°; the motion for 26 avadhis (= 6,4 days) is −19;18,44°.
F. 4r. Table 1 of the Anonymous of 1741.
F. 4r. Table 2 of the Anonymous of 1741.
F. 4v. Table 3 of the Anonymous of 1741.
F. 4v. Table 4 of the Anonymous of 1741.
F. 5r. Table 5 of the Anonymous of 1741.
F. 5r. Table 6 of the Anonymous of 1741.
F. 5v. Table 7 of the Anonymous of 1741.
F. 5v. Table 8 of the Anonymous of 1741.
F. 6r. Table 9 of the Anonymous of 1741.
F. 6r. Table 10 of the Anonymous of 1741.
F. 6v. Table 11 of the Anonymous of 1741 for 0 to 37 periods.
F. 7r. Table 12 of the Anonymous of 1741 for 0 to 41 periods.
F. 7v. Table 13 of the Anonymous of 1741 for 0 to 43 periods.
Ff. 8r–10v. Table 16 of the Anonymous of 1741 for 0 to 27 horizontal, and for 0 to 59 vertical.
Ff. 11r–13v. Table 17 of the Anonymous of 1741 for 0 to 27 horizontal, and for 0 to 59 vertical.
Ff. 14r–16v. Table 18 of the Anonymous of 1741 for 0 to 29 horizontal, and for 0 to 59 vertical.

4771 (Smith Indic 91) ff. 1–8. On f. 1r, at the bottom, is written in European numerals: 6584. On f. 8v, at the bottom, is written: śloka 200.
Ff. 1r–1v. Table 1 of the *Mahādevī*.
Ff. 2r–2v. Table 2 of the *Mahādevī*.
Ff. 3r–3v. Table 3 of the *Mahādevī*.
Ff. 4r–4v. Table 4 of the *Mahādevī*.
Ff. 5r–5v. Table 5 of the *Mahādevī*.
Ff. 6r–6v. Table 6 of the *Mahādevī*.
Ff. 7r–7v. Table 7 of the *Mahādevī*.
Ff. 8r–8v. Table 8 of the *Mahādevī*.

4772 (Smith Indic 193) ff. 49–75.

Ff. 49r–60v. Table 12 of the *Mahādevī* (N = 12 to 59); columns A, B, D, and F, and function G.
Ff. 61r–75v. Table 13 of the *Mahādevī* (N = 0 to 59); columns A, B, D, and F, and function G.

4773 (Smith Indic 147) ff. 31–60 and unnumbered f. See 4769.

4777 (Smith Indic 33) 34 ff. in two parts: (a) consists of ff. 1–30, (b) of ff. 1–4. On (a) f. 1r is written: atha grahaprabodhasāriṇi prārambhaḥ; on (b) f. 4r is written: idaṃ pustakaṃ govindaśarmaṇā nāsikara ity upanāmnā likhitaṃ // śake 1793 vaiśākha vadya 2 śanau. This date corresponds to Saturday, 6 May 1871. On (a) f. 30v, at the bottom, is written: ślo 750.

(a) f. 1v. Table 1 of the *Grahaprabodhasāriṇī*.
Ff. 2r–2v. Table 2 of the *Grahaprabodhasāriṇī*.
Ff. 3r–3v. Table 3 of the *Grahaprabodhasāriṇī*.
Ff. 4r–4v. Table 4 of the *Grahaprabodhasāriṇī*.
Ff. 5r–5v. Table 5 of the *Grahaprabodhasāriṇī*.
Ff. 6r–6v. Table 6 of the *Grahaprabodhasāriṇī*.
Ff. 7r–7v. Table 7 of the *Grahaprabodhasāriṇī*.
Ff. 8r–8v. Table 8 of the *Grahaprabodhasāriṇī*.
Ff. 9r–9v. Table 9 of the *Grahaprabodhasāriṇī*.
Ff. 10r–10v. Table 10 of the *Grahaprabodhasāriṇī*.
Ff. 11r–12v. Table 11 of the *Grahaprabodhasāriṇī*.
Ff. 13r–14v. Table 12 of the *Grahaprabodhasāriṇī*.
Ff. 15r–16v. Table 13 of the *Grahaprabodhasāriṇī*.
Ff. 17r–18v. Table 14 of the *Grahaprabodhasāriṇī*.
Ff. 19r–20v. Table 15 of the *Grahaprabodhasāriṇī*.
Ff. 21r–22v. Table 16 of the *Grahaprabodhasāriṇī*.
Ff. 23r–24v. Table 17 of the *Grahaprabodhasāriṇī*.
Ff. 25r–25v. Table 18 of the *Grahaprabodhasāriṇī*.
Ff. 26r–26v. Table 19 of the *Grahaprabodhasāriṇī*.
Ff. 27r–27v. Table 20 of the *Grahaprabodhasāriṇī*.
Ff. 28r–28v. Table 21 of the *Grahaprabodhasāriṇī*.
Ff. 29r–29v. Table 22 of the *Grahaprabodhasāriṇī*.
F. 30r. Table 23 of the *Grahaprabodhasāriṇī*.
F. 30v. Table 24 of the *Grahaprabodhasāriṇī*.

(b) ff. 1r–4r. Explanatory text of Yādava.
F. 4r. After the colophon indicating the date of the manuscript are several lines concerning the elongation from the Sun necessary for the occurrence of various Greek-letter phenomena:

	First station	*Second station*	*First visibility*	*Last visibility*
Mars	163°	197°	18°	332°[342]
Mercury	145	215	50 West, 205 East	155W, 310E
Jupiter	125	235	14	346
Venus	167	193	20[24] W, 183E	177W, 336E
Saturn	113	247	17	343

F. 4v. Blank.

4778 (Smith Indic 30) 18 ff. On f. 18v, at the top, is written: śloka 540; on f. 18v, at the bottom, is written: saṃvat 1806 varṣe posa va di 10 ravau śrīmāṃḍavimadhye josī hīrajī sva ātmārthe. This date corresponds to Sunday, 21 January 1750 in the Julian calendar.
Ff. 1v–3v. Text of the *Pañcāṅgavidyādharī*.
F. 2r. Parameters: tithikendra 27;59,33; nakṣatrakendra 27;13,48; and yogakendra 29;16,1.
F. 3r. Table 1 of the *Pañcāṅgavidyādharī*.
F. 3v. Table 2 of the *Pañcāṅgavidyādharī* for 1 to 12 zodiacal signs. In the margin is written: brahmapakṣe abdape joyyā ayanāṃśā 19.
F. 3v. Table 3 of the *Pañcāṅgavidyādharī*.
F. 4r. Table 4 of the *Pañcāṅgavidyādharī* for 1 to 27 nakṣatras.
Ff. 4v–5r. Table 5 of the *Pañcāṅgavidyādharī* for 1 to 38 periods.
Ff. 5v–6r. Table 6 of the *Pañcāṅgavidyādharī* for 1 to 41 periods.
Ff. 6v–7r. Table 7 of the *Pañcāṅgavidyādharī* for 1 to 44 periods.
F. 7v. Table 8 of the *Pañcāṅgavidyādharī*.
F. 8r. Table called saṃkrāntisvarūpa; ascribed to the *Muhūrtacintāmaṇi* (of Rāma). See the third adhyāya of that work.
Ff. 8v-12r. Table 9 of the *Pañcāṅgavidyādharī* for 0 to 27 horizontal, and for 0 to 59 vertical.
Ff. 12v–15r. Table 10 of the *Pañcāṅgavidyādharī* for 0 to 27 horizontal, and for 0 to 59 vertical.
Ff. 15v–18v. Table 11 of the *Pañcāṅgavidyādharī* for 0 to 29 horizontal, and for 0 to 59 vertical.

4798, 4799, 4955 (U. Penn. 1913) 4 ff. This number contains three manuscripts: (a) of 2 ff., (b) of 1 f., and (c) of 1 f. It is unclear whether Poleman intended his numbers to refer to these. On f. (a) 1r is written: śrīgaṇeśāya namaḥ // śrīsarasvatyai namaḥ // aṃśādīni ravimaṃdaphalāni likhyaṃte // akṣāṃśāḥ 25;26,37,30 // śaka 1798; on f. 2r is written the same thing, except that iṃduphalāni is substituted for ravimaṃdaphalāni. So the manuscript was written in A.D. 1876 in a place of 25;26,37,30° North latitude.

(a) ff. 1r–1v (4798). Table 19 of Anonymous of 1520 for 0° to 90°.

(a) ff. 2r–2v. (4799) Table 20 of Anonymous of 1520 for 0° to 90°.

(b) ff. 1r–1v (4955). Table of accumulated rising-times of the degrees of the zodiac measured in ghaṭikās, for 0° to 29° horizontal, 0 to 11 signs vertical. The vernal equinox is placed at Pisces 8°.

(c) f. 1r. Table of 10 attributes of the 16 figures used in ramalaśāstra (geomancy).

(c) f. 1v. Blank.

4799 (U. Penn. 1913(a) ff. 2r–2v). See 4788.

4800 (Smith Indic 129 A ff. 12r–21v). See 4801.

4801, 4800, 4802, 5188, 5187 (Smith Indic 129) 63 ff. This number consists of a cover folio (f. I), and two manuscripts: A containing ff. 1–33, and B containing ff. 2–30. On f. Ir is written: aṃkacālaprāraṃbhaḥ titheḥ. On A f. 33v, at the top, is written: śloka 700.

A ff. 1r–11v (4801). Table 1 of the *Tithicintāmaṇi* for 0 to 400 tithis.
F. 11v. Table la (of the *Tithicintāmaṇi*).
F. 12r (4800). Blank except for the title: aṃkacāla nakṣatrasya.
Ff. 12v–21v. Table 2 of the *Tithicintāmaṇi* for 0 to 390 nakṣatras.
F. 22r (4802). Blank except for the title: aṃkacāla yogasya.
Ff. 22v–32v. Table 3 of the *Tithicintāmaṇi* for 0 to 413 yogas.
F. 33r. Blank.

F. 33v. Blank except for the colophon: aṃkacāla yogasya saṃpūrṇa.

B f. 2r (5188). Table 1 of the *Makaranda* for Śaka 1656 to 1768 (A.D. 1734 to 1846); the order of the columns is 1, 2, 4, 3, and column 4 is entitled kendra instead of vallī.
F. 2r. Table 2 of the *Makaranda* for 1 to 16 years.
Ff. 2v–4r. Table 4 of the *Makaranda* for 0 to 59 horizontal, and, in intervals of 6, for 0 to 54 vertical.
Ff. 4v–7v. Table 4a (of the *Makaranda*) for 1 to 404 tithis. The following entries are not filled in: 117–176 on f. 5v; and 235–354 on ff. 6v–7r.
F. 8r. Table 5 of the *Makaranda* for Śaka 1664 to 1832 (A.D. 1742 to 1910); the same arrangement holds as for table 1 of the *Makaranda* in this manuscript.
F. 8r. Table 6 of the *Makaranda* for 1 to 24 years.
Ff. 8v–10r. Table 8 of the *Makaranda* for 0 to 59 horizontal, and, in intervals of 6, for 0 to 54 vertical.
Ff. 10v–13v. Skeleton of a table for mean nakṣatras similar to table 4a for mean tithis for 1 to 414 nakṣatras; but nothing filled in.
F. 14r. Table 9 of the *Makaranda* for Śaka 1664 to 1832 (A.D. 1742 to 1910); the same arrangement holds as for table 1 of the *Makaranda* in this manuscript.
F. 14r. Table 10 of the Makaranda for 1 to 24 years.
Ff. 14v–16r. Skeleton of table 12 of the *Makaranda* for 0 to 59 horizontal, and, in intervals of 6, for 0 to 54 vertical; nothing filled in.
Ff. 16v–19v. Skeleton of a table for mean yogas similar to table 4a for mean tithis for 1 to 403 yogas; nothing filled in.
F. 20r. Table 13 of the *Makaranda* in intervals of 16 years for Śaka 1656 to 1784 (A.D. 1734 to 1862).
F. 20r. Table 14 of the *Makaranda* for 1 to 16 years.
F. 20r. Table 16c (of the *Makaranda*) for 1 to 28 nakṣatras.
F. 20v. Table 15 of the *Makaranda,* with an added column of cālanas according to the school of Gaṇeśa (gaṇeśadaivajñapakṣaghaṭanārthaṃ).
F. 21r. Table of cālanas for table 16c, also according to the school of Gaṇeśa.
F. 21r. A nakṣatracaraṇapraveśapattra for the first caraṇa of 1 to 27 nakṣatras.
Ff. 21v–22r (5187). Blank.
F. 22v. Saṃkramaṇapraveśakṣepakāḥ.
F. 22v. Āryapakṣasaṃkrāntisaṃskāraḥ.
Ff. 22v–23r. Aśvinedi (*sic!*) nakṣatrapraveśakṣepakāḥ.
F. 23r. Yearly cālanas for the following functions: Lord of the Year, Epact, tithidhruva, nakṣatrayogadhruva, nakṣatradhruva, yogadhruva, tithimadhyakendra, nakṣatramadhyakendra, yogamadhyakendra, tithispaṣṭakendra, nakṣatraspaṣṭakendra, yogaspaṣṭakendra, tithibhoga, nakṣatrabhoga, yogabhoga, and bhabhoga. *Cf.* table 8 (of the *Tithicintāmaṇi*).
F. 23v. Kṣepakas of the same functions according to the: *Saurapakṣa, Āryapakṣa, Brahmapakṣa,* and *Keśavapakṣa.* To the right is written: idaṃ gaṇeśadaivajñalaghucintāmaṇyupakaraṇyo ṛṇadhanātmakaṃ cālanaṃ vidheyaṃ tattatpakṣīyopakaraṇāni syuḥ.
F. 24r. Blank except for the title: saṃkrāntimahānakṣatrakṣepakāḥ.
F. 24v. Saṃkramaṇapraveśakṣepakāḥ for Aries to Pisces.
F. 24v. Saṃkrāntisaṃskāro āryapakṣe dhana-ṛṇātmakaṃ palāni for Aries to Pisces.
F. 25r. Mahānakṣatrapraveśakṣepakāḥ for 1 to 28 nakṣatras.
F. 25v. Contains same table as is on f. 23r.
Ff. 26r. Blank.
Ff. 26v–27r. Scribblings.
Ff. 27v–28r. Blank.
Ff. 28v–30r. Fragment of an unidentified astrological text.
F. 30v. End of a text on divination by crows, in a different hand.

4802 (Smith Indic 129A ff. 22–33v). See 4801.

4805 (Smith Indic 100 ff. 8r–9r). See 4904.

4807 (Smith Indic 27) 1 f. On f. 1v, at the bottom is written: ity upakaraṇayaṃtrapatraṃ likhitaṃ malūkacandrarṣiṇā rādhanapure.
F. 1r–1v. Tables 1 and 2 of the Anonymous of 1598.

4814 (U. Penn. 1859) 12 ff. On f. 12v is written: saṃ 1854 vai śu 3 śanau li cīreśvara mahājana. So the manuscript was written by Cīreśvara Mahājana on Saturday, 29 April 1797.
Ff. 1v–5r. Table 1 of the *Tithicintāmaṇi* for 0 to 400 tithis.
Ff. 5v–8r. Table 2 of the *Tithicintāmaṇi* for 0 to 390 nakṣatras.
Ff. 8r–12r. Table 3 of the *Tithicintāmaṇi* for 0 to 419 yogas.
F. 12v. Table 4 of the *Tithicintāmaṇi* for 1 to 27 nakṣatras.
F. 12v. Table 5 of the *Tithicintāmaṇi* for 1 to 12 zodiacal signs.

4819 (Harvard 59) 7 ff. On f. 7v is written: saṃ 1866 caitraśukla 15 bhṛgu / śrīmadgrahalāghavīyamadhyamaspaṣṭārkasāriṇī josī khemyanātha likhitā paṭhanārthaṃ josījī śrīcaṃbanarāmajī. The date corresponds to Thursday, 31 March 1809.
Ff. 1r–7v. Table 1 of the *Grahalāghavīyamadhyamaspaṣṭārkasāriṇī* for 0 to 365 days.
F. 7v. Brief explanatory text.

4821 (Smith Indic 100 ff. 12r–12v). See 4904.

4822 (Smith Indic 100 ff. 13r–20v). See 4904.

4823 (Smith Indic 190 ff. 13–17v). See 4717.

4824 (Smith Indic 190 ff. 8–12v). See 4717.

4825 (Smith Indic 58) ff. 1–15. On f. 15v is written: śloka 400. On f. 1r is written: candrārkikoṣṭakāni ślo 372; on f. 15v: iti candrārki samāptaḥ.
Ff. 1r–5v. Table 1 of the *Candrārkī* for 0 to 365 days.
Ff. 6r–12v. Table 2 of the *Candrārkī* for 0 to 355 days.
Ff. 13r–15v. Table 4 of the *Candrārkī* for 0° to 359°.

4826 (Harvard 934) 14 ff.
F. 1r. Blank.
Ff. 1v–5r. Table 1 of the *Candrārkī* for 0 to 365 days.
Ff. 5v–9r. Table 2 of the *Candrārkī* for 0 to 365 days.
F. 9v. Table 3 of the *Candrārkī* for 1 to 60 ghaṭikās.
Ff. 10r–14r. Table 4 of the *Candrārkī* for 0° to 359°.
F. 14v. Blank.

4827 (Smith Indic 180) ff. 1–2. On f. 1r is written: ślo 48, and, in European numerals: 6593. On f. 2v, at the bottom, is written: iti śrīcandrārkīnāṃ upakaranāṃ samāptaḥ // saṃvat 1829 varṣe śāke 1694 pra phāguna su di 7 ravau śrīrādhanapure likhitaṃ. This date corresponds to Sunday, 28 February 1773. The scribe was probably the Malūkacandra who wrote 4807.
Ff. 1r–2v. Table 5 of the *Candrārkī*.

4853, 4857, 4862 (Smith Indic 166) 31 ff. On f. 1r, to the left, is written, in European numerals: 6414.
Ff. 1r–1v (4853). Table 1 of the *Makaranda* for Śaka 1512 to 1772 (A.D. 1590 to 1850).
F. 2r. Table 2 of the *Makaranda* for 1 to 16 years.
F. 2v. Blank.
Ff. 3r–5v (4857). Table 4 of the *Makaranda* for 0 to 59 horizontal and 0 to 9 vertical, and for 0 to 32 horizontal and 10 to 19 vertical. (See below, ff. 10r–18v.)
F. 6r. Blank.
Ff. 6v–7r. Notes.
F. 7r. Table 52 (of the *Makaranda?*).
F. 8r. Table 5 of the *Makaranda* for Śaka 1616 to 1832 (A.D. 1694 to 1910).
F. 8r. Table 6 of the *Makaranda* for 1 to 24 years.
F. 8v. Short note.
F. 9r. Table 15 of the *Makaranda* for 1 to 12 zodiacal signs.
Ff. 9r–9v. Table 16 of the *Makaranda* for 1 to 27 nakṣatras.
F. 9v. Table 37 (of the *Makaranda*?).
Ff. 10r–18v. Table 4 of the *Makaranda* for 33 to 59 horizontal and 10 to 19 vertical, and for 0 to 59 horizontal and 20 to 59 vertical (see above, ff. 3r–5v).
F. 19r (4862). Table 9 of the *Makaranda* (without the column of vallīs) for Śaka 1616 to 1784 (A.D. 1694 to 1862).
Ff. 19r–18v (*sic!*). Table 6 of the *Makaranda* (without the column of vallīs) for 1 to 24 years.
F. 19v. Table 7 of the *Makaranda* for 0 to 14 sidereal months.
Ff. 20r–31v. Table 8 of the *Makaranda* for 0 to 59 horizontal, and for 0 to 59 vertical.

4854 (U. Penn. 709) 12 ff.
Ff. 1v–5v. Table 1 of the *Tithicintāmaṇi* for 0 to 402 tithis.
Ff. 6r–9r. Table 2 of the *Tithicintāmaṇi* for 0 to 390 nakṣatras.
Ff. 9v–12v. Table 3 of the *Tithicintāmaṇi* for 0 to 357 yogas.

4855, 4861, 4892, 4859, 5125, 4951, 4943 (Smith Indic 138) ff. 1–28. On f. 1r is written: atha tithicintāma; a later hand has added: ṇikoṣṭakāni, and also: patra 27 ślo 375. At the bottom of the folio, in European numerals, is written: 6534.
Ff. 1v–8v (4855). Table 1 of the *Tithicintāmaṇi* for 0 to 400 tithis.
F. 9r. Blank.
Ff. 9v–14r (4861). Table 2 of the *Tithicintāmaṇi* for 0 to 390 nakṣatras.
F. 14v. Blank.
Ff. 15r–23v (4892). Table 3 of the *Tithicintāmaṇi* for 0 to 420 yogas.
Ff. 24r–26v (4859). Table 1 of the Anonymous of 1461 for 1 to 366 days.
F. 26v. Table 2 of the Anonymous of 1461.
F. 26v. Table 3 of the Anonymous of 1461 to Lunar node (1).
F. 26v. Table 4 of the Anonymous of 1461 to Jyeṣṭhā.
F. 27r (5125). Table 4 of the *Tithicintāmaṇi* for 1 to 27 nakṣatras.
F. 27v (4951). Table 5 of the *Tithicintāmaṇi* for 1 to 12 zodiacal signs.
F. 27v. Table 3 of the Anonymous of 1461, complete. There is an added note: varṣasya sāvayavadināni likhyaṃte 365;15,31,30 / saptaviṃśatibhāgātmako aharagaṇaḥ (read: 'hargaṇaḥ) 13;31,41,10 avadhist (read: avadhiḥ syāt) / cālanaṃ 6;31,41,10.

If a sidereal year equals 6,5;15,31,30 days, then 1/27 of a year is indeed 13;31,41,10 days; but an avadhi is 14 days in the usage of all other texts available at present. The cālana appears simply to be 13;31,41,10 modulo 7.
F. 28 (4943) is missing.

4856 (U. Penn. 1799) 27 ff. There are 4 sections with separate folio-numeration: (a) ff. 1–6; (b) ff. 1–6; (c) ff. 1–7; and (d) ff. 1–8. On (a) f. 1r is written: atha tithicintāmaṇisāraṇiḥ. On (d) f. 8v is written: iti tithicintāmaṇisāraṇiḥ.

(a) ff. 1v–6r. Table 1 of the *Tithicintāmaṇi* for 0 to 400 tithis.
F. 6v. Blank.

(b) ff. 1r–6v. Table 2 of the *Tithicintāmaṇi* for 0 to 390 nakṣatras.

(c) ff. 1r–6r. Table 3 of the *Tithicintāmaṇi* for 0 to 400 yogas.
F. 6v. Blank.

F. 7r. Table 6 of the *Tithicintāmaṇi.*
Ff. 7r–7v. Table 7 of the *Tithicintāmaṇi.*

(d) f. 1r. Table 8 (of the *Tithicintāmaṇi?*).
F. 1v. Table 4 of the *Tithicintāmaṇi* for 1 to 27 nakṣatras.
F. 1v. Table 5 of the *Tithicintāmaṇi* for 1 to 12 zodiacal signs.
Ff. 2r–2v. Table 9 (of the *Tithicintāmaṇi?*) for 1 to 10 periods of 46 years.
Ff. 3r–8r. Table 10 (of the *Tithicintāmaṇi?*) for 0 to 46 years.

4857 (Smith Indic 166 ff. 3r–18v). See 4853.

4858 (Smith Indic 95) 1 f. On f. 1v, at the bottom, is written: likhitaṃ puṣkarṇājñātīya paṇiā josī śrī 5 dośā mukadajī // saṃvat 1916 nā śāke 1781 prå āśu śu di 7 dine. This date corresponds to 3 October 1859. On f. 1v, at the top, is written: ślo 30, and, at the bottom, in European numerals: 6574.
Ff. 1r–1v. A ḍinamānapattra; the heading is: atha dinamānapatram//'yanāṃśā 22 noche//gurjjarakachasaurāṣṭropāṃcālesiṃdhuparvatejñe / yam. The maximum length of daylight is 33;34,0 ghaṭikās at Gemini 8°. Also given are the rising-times of the signs in palas:

Aries, Virgo, Libra, Pisces	278
Taurus, Leo, Scorpio, Aquarius	299
Gemini, Cancer, Sagittarius, Capricorn	323

4859 (Smith Indic 138 ff. 24r–26v). See 4855.

4860 (U. Penn. 1895) 2 ff. On f. 2r, in the upper margin, is written the equinoctial noon shadow (5;28) for rīvāṃda (Rewah); and, in the right-hand margin, the rising-times in palas of the zodiacal signs at rīvāgrāma (Rewah). On f. 2v are written distances to the following places: positive indicates South, negative North.

Brahmāvarta	−28 yojanas
Kurukṣetra	−28 yojanas
Puṇe (Poona)	+28 yojanas
Nāśika	+27 yojanas
Vāi (Wai)	+24 yojanas

Ff. 1r–2r. Geographical Table.

4861 (Smith Indic 138 ff. 9v–14r). See 4855.

4862 (Smith Indic 166 ff. 19r–31v). See 4853.

4863 (Smith Indic 81) 3 ff. On f. 1r, in the upper left, is written: śloka 300, and, in the lower right in European numerals: 6594.
Ff. 1r–3v. Table 1 of the Anonymous of 1578.

4864 (Smith Indic 84) 2 ff. Written by the same scribe that wrote 4865.
Ff. 1r–1v. Tables of true longitudes of the five planets for Saṃ. 1857, Śaka 1722 (A.D. 1800) computed from the *Mahādevī*; only columns A, B, and D, and function G are given. The mean motions of the planets at the beginning of the year are given in the margins (in terms of 6° or *Mahādevī* tables):

Mars	44;5,5,9
Mercury	27;27,29,42
Jupiter	11;45,5,24
Venus	48;29,46,51
Saturn	17;4,46,56

Ff. 2r–2v. Tables of true longitudes of the five planets for Saṃ. 1859, Śaka 1724 (A.D. 1802) computed from the *Mahādevī*, set up as are those on ff. 1r–1v. The mean motions till the beginning of the year are:

Mars	47;53,8,15
Mercury	45;42,36,2
Jupiter	21;52,7,34
Venus	3;33,44,47
Saturn	21;9,4,6

4865 (Smith Indic 85) 2 ff. On f. 2v, in the upper left-hand corner, is written: śloka 120, and in the lower right-hand corner in European numerals: 6587. Written by the same scribe that wrote 4864.
Ff. 1r–1v. Tables of true longitudes of the five planets for Saṃ. 1865, Śaka 1730 (A.D. 1808) computed from the Anonymous of 1704; columns A, B, and M, and function G are given. The positions of the planets in terms of the tables (N) are thus enumerated:

Mars	25	Venus	104
Mercury	12	Saturn	45
Jupiter	21	Lunar Node	11

Ff. 2r–2v. Tables of true longitudes of the five planets for Saṃ. 1866, Śaka 1731 (A.D. 1809) computed from the Anonymous of 1704, set up as are those on ff. 1r–1v. The values of N are:

Mars	26	Venus	105
Mercury	13	Saturn	46
Jupiter	22	Lunar Node	12

4866 (Smith Indic 101) 8 ff. On f. 1r, in the upper margin, is written in European numerals: 1657. On f. 8v, in the upper margin, is written: śloka 225.
Ff. 1r–1v. Table 1 of the *Mahādevī.*
Ff. 2r–2v. Table 2 of the *Mahādevī.*
Ff. 3r–3v. Table 3 of the *Mahādevī.*
Ff. 4r–4v. Table 4 of the *Mahādevī.*
Ff. 5r–5v. Table 5 of the *Mahādevī.*
Ff. 6r–6v. Table 6 of the *Mahādevī.*
Ff. 7r–7v. Table 7 of the *Mahādevī.*
Ff. 8r–8v. Table 8 of the *Mahādevī.*

4867 (Smith Indic 105) 46 ff. Each section of this manuscript is numbered separately. On (a) f. 1r, in the upper left-hand corner, is written: 13, and, in the upper margin: śloka 1000. On (a) f. 9r, in the lower right-hand corner, is written in European numerals: 6572.

(a) ff. 1–30. Table 9 of the *Mahādevī* (N = 0 to 59).

(b) ff. 2–9. Table 10 of the *Mahādevī* (N = 2 to 17).

(c) ff. 1r–1v. Table 1 of the *Mahādevī*.
Ff. 2r–2v. Table 2 of the *Mahādevī*.
Ff. 3r–3v. Table 3 of the *Mahādevī*.
Ff. 4r–4v. Table 4 of the *Mahādevī*.
Ff. 5r–5v. Table 5 of the *Mahādevī*.
Ff. 6r–6v. Table 6 of the *Mahādevī*.
Ff. 7r–7v. Table 7 of the *Mahādevī*.
Ff. 9r–9v. (*sic*). Table 8 of the *Mahādevī*.

4868 (Smith Indic 140) 177 ff. A composite of parts of 3 manuscripts: A has 18 ff., B 8 ff., C 150 ff., and D 1 f.; B and C are from the same manuscript. On A f. 1r, in the left-hand margin, is written: Author—Mahadev / Title—Mahadev's Koshtak / abt. 300 years ago. On D f. 1v, to the left, is written: 31, and to the right: śloka 3000.

A ff. 1r–18v. Table 9 of the *Mahādevī* (N = 0 to 35).

B ff. 4r–4v. Table 1 of the *Mahādevī*.
Ff. 5r–5v. Table 2 of the *Mahādevī*.
Ff. 6r–6v. Table 3 of the *Mahādevī*.
Ff. 7r–7v. Table 4 of the *Mahādevī*.
Ff. 8r–8v. Table 5 of the *Mahādevī*.
Ff. 9r–9v. Table 6 of the *Mahādevī*.
Ff. 10r–10v. Table 7 of the *Mahādevī*.
Ff. 11r–11v. Table 8 of the *Mahādevī*.

C ff. 1r–30v. Table 9 of the *Mahādevī* (N = 0 to 59).
Ff. 31r–60v. Table 10 of the *Mahādevī* (N = 0 to 59).
Ff. 61r–90v. Table 11 of the *Mahādevī* (N = 0 to 59). Facsimile of f. 61 in *Proc. Amer. Philos. Soc.* 111, 1967, 76.
Ff. 91r–120v. Table 12 of the *Mahādevī* (N = 0 to 59).
Ff. 121r–150v. Table 13 of the *Mahādevī* (N = 0 to 59).

D f. 1r. Table 1 of the Anonymous III of 1578 for Śaka 1500 to 1700 (A.D. 1578 to 1778).
F. 1r. Table 2 of the Anonymous III of 1578 for 1 to 27 avadhis.
F. 1v. Blank.

4869 (Smith Indic 146) 100 ff. The ff. are numbered separately for each planet by the original scribe, consecutively throughout the manuscript by a later scribe, whom I follow. This later scribe has also written on f. 1r: jagadbhūṣaṇasaṃjñakasya jyotirgraṃthasya koṣṭhakāni. On f. 1r, at the top, is written: śloka 3000. On f. 96r, in the upper left-hand corner, is written: jagat bhuṣaṇa.
Ff. 1v–21r. Table 1 of the *Jagadbhūṣaṇa* (N = 0 to 78).
F. 21v. Blank.
Ff. 22r–33r. Table 2 of the *Jagadbhūṣaṇa* (N = 0 to 45).
F. 33v. Blank.
Ff. 34r–47v. Table 3 of the *Jagadbhūṣaṇa* (N = 0 to 82).
Ff. 48r–85v. Table 4 of the *Jagadbhūṣaṇa* (N = 0 to 226).
Ff. 86r–95v. Table 5 of the *Jagadbhūṣaṇa* (N = 0 to 58).
F. 96r. Table 6 of the *Jagadbhūṣaṇa* for 0 to 88 years.
Ff. 96v–97r. Table 7 of the *Jagadbhūṣaṇa* for 0 to 121 years.
Ff. 97r–97v. Table 8 of the *Jagadbhūṣaṇa* for 0 to 121 years.
Ff. 97v–98r. Table 9 of the *Jagadbhūṣaṇa* for 0 to 43 years.
F. 98r. Table 10 of the *Jagadbhūṣaṇa* for 1 to 27 avadhis.
F. 98r. Table 11 of the *Jagadbhūṣaṇa* for 1 to 27 avadhis.
F. 98r. Table 12 of the *Jagadbhūṣaṇa* for 1 to 13 days.
F. 98v. Table 13 of the *Jagadbhūṣaṇa* for 1 to 13 days.
F. 98v. Table 14 of the *Jagadbhūṣaṇa* for 1 to 60 ghaṭikās.
F. 99r. Table 15 of the *Jagadbhūṣaṇa* for 0 to 30 intervals of 3° each.
Ff. 99r–99v. Table 16 of the *Jagadbhūṣaṇa* for 1 to 27 avadhis.
F. 99v. Table 17 of the *Jagadbhūṣaṇa* for 1 to 27 avadhis.
F. 99v. Table 18 of the *Jagadbhūṣaṇa* for 1 to 27 avadhis.
F. 100r. Table 19 of the *Jagadbhūṣaṇa* for 0 to 92 years.
F. 100v. Table 20 of the *Jagadbhūṣaṇa* for 1 to 27 avadhis.
F. 100v. Table 21 of the *Jagadbhūṣaṇa* for 1 to 12 zodiacal signs and 1 to 27 nakṣatras.

4870 (Smith Indic 191) 2 ff. On f. 1v, in the left-hand margin, is written: śloka 400.
Ff. 1r–2v. Tables of the true longitudes of Mars, Mercury, Jupiter, and Venus for Saṃ. 1842 (A.D. 1786) computed from the *Mahādevī*; only columns A, B, and C, and function G are given. The mean positions of the planets at the beginning of the year in terms of *Mahādevī*-tables are given as:

Mars	45;34,41,48
Mercury	10;34,12,6
Jupiter	(55)
Venus	25;30,2,19,42

4876 (Smith Indic 43) ff. 1–13. On f. 13v is written: śloka 300.
F. 1r. Table 12 of the *Brahmatulyasāraṇī*.
Ff. 1r–1v. Table 13 of the *Brahmatulyasāraṇī*.
Ff. 1v–2r. Table 10 of the *Brahmatulyasāraṇī*.
Ff. 2r–2v. Table 11 of the *Brahmatulyasāraṇī*.
Ff. 2v–3r. Table 14 of the *Brahmatulyasāraṇī*.
Ff. 3r–3v. Table 16 of the *Brahmatulyasāraṇī*.
Ff. 4r–4v. Table 18 of the *Brahmatulyasāraṇī*.
Ff. 4v–5r. Table 20 of the *Brahmatulyasāraṇī*.
Ff. 5r–5v. Table 22 of the *Brahmatulyasāraṇī*.

Ff. 6r–7v. Table 15 of the *Brahmatulyasāraṇī*.
Ff. 8r–9r. Table 17 of the *Brahmatulyasāraṇī*.
Ff. 9v–10v. Table 19 of the *Brahmatulyasāraṇī*.
Ff. 11r–12r. Table 21 of the *Brahmatulyasāraṇī*.
Ff. 12v–13v. Table 23 of the *Brahmatulyasāraṇī*.

4877, 4675 (Smith Indic 185) 55 ff.
F. 1r (4877). Table 1 of the *Makaranda* for Śaka 1480 to 1528 and 1592 to 1624 (A.D. 1558 to 1606 and 1670 to 1702); a different scribe has added the entries for Śaka 1672 (A.D. 1750).
F. 1r. Table 2 of the *Makaranda* for 1 to 16 years.
F. 1v. Table 2 of the *Makaranda* for 1 to 16 years repeated.
F. 2r. Table 3 of the *Makaranda* for 0 to 26 pakṣas.
Ff. 2v–3v. Table 4 of the *Makaranda* for 0 to 59 horizontal, and, in intervals of 6, for 0 to 54 vertical.
F. 4r. Table 7 of the *Makaranda* for 1 to 14 sidereal months.
Ff. 4v–5v. Table 8 of the *Makaranda* for 0 to 59 horizontal, and, in intervals of 6, for 0 to 54 vertical.
F. 6r. Table 5 of the *Makaranda* for Śaka 1400 to 1576 (A.D. 1478 to 1654).
F. 6r. Table 6 of the *Makaranda* for 1 to 24 years.
F. 6v. Table 9 of the *Makaranda* for Śaka 1400 to 1496 (A.D. 1478 to 1574).
F. 6v. Table 10 of the *Makaranda* for 1 to 24 years.
F. 7r. Table 11 of the *Makaranda* for 0 to 14.
Ff. 7v–8v. Table 12 of the *Makaranda* for 0 to 59 horizontal, and, in intervals of 6, for 0 to 54 vertical.
F. 9r. Table 13 of the *Makaranda* for Śaka 1400 and 1448 to 1520 (A.D. 1478 and 1526 to 1598).
F. 9r. Table 14 of the *Makaranda* for 1 to 24 years.
F. 9r. Table 15 of the *Makaranda* for 1 to 12 zodiacal signs.
F. 9r. Table 16 of the *Makaranda* for 1 to 27 nakṣatras.
Ff. 9v–10r. Table 38 of the *Makaranda* for 0° to 29° horizontally, and for 1 to 12 signs vertically. The maximum length of daylight is 34;4 ghaṭikās at Cancer 0°.
Ff. 10v–11r. Table 17 of the *Makaranda* for Śaka 1400 to 1571 (A.D. 1478 to 1649), for 1 to 57 years, and for 1 to 23 pakṣas.
Ff. 11v–12v. Table 18 of the *Makaranda*.
Ff. 12r–12v. Table 19 of the *Makaranda*.
Ff. 12v–13v. Table 20 of the *Makaranda*.
Ff. 13v–14r. Table 21 of the *Makaranda*.
Ff. 14v–15r. Table 22 of the *Makaranda*.
Ff. 15r–16r. Table 23 of the *Makaranda*.
Ff. 16r–16v. Table 24 of the *Makaranda*.
Ff. 16v–17r. Table 25 of the *Makaranda*.
Ff. 17r–18r. Table 26 of the *Makaranda*.
F. 18r. Table 26a (of the *Makaranda?*).
F. 18r. Table 26b (of the *Makaranda?*) for Leo to Pisces.
Ff. 18v–19r. Table 27 of the *Makaranda*, combined in 1 table, for 1° to 90°.
Ff. 19v–20v. Table 28 of the *Makaranda* for 1° to 89°.
Ff. 20v–22v. Table 29 of the *Makaranda* for 1° to 179°; none of the additional parameters is given.
Ff. 23r–25v. Table 30 of the *Makaranda* for 1° to 180°; only the apogee is given extra.
Ff. 25v–28r. Table 31 of the *Makaranda* for 1° to 180°; none of the additional parameters is given.
Ff. 28r–30v. Table 32 of the *Makaranda* for 1° to 180°; only the apogee is given extra.
Ff. 30v–32r. Table 33 of the *Makaranda* for 1° to 180°; only the apogee is given extra.
Ff. 32v–33r. Table 38a (of the *Makaranda?*) for 1 to 60.
F. 33r. Table 48 (of the *Makaranda?*) for 0;56 to 1;6 days.
F. 33v. Table 49 (of the *Makaranda?*) for 1 to 25 (= 31 to 55).
F. 33v. Table 50 (of the *Makaranda?*) for 1 to 12 zodiacal signs.
F. 33v. Table 37 (of the *Makaranda?*).
F. 33v. Table 51 (of the *Makaranda?*).
Ff. 34r–35r. Table similar to table 41 (of the *Makaranda?*), but giving the beginnings and ends of the nakṣatras rather than of the nakṣatracaraṇas.
F. 35v. Table 52 (of the *Makaranda?*).
Ff. 36r–37v. Table 36 (of the *Makaranda?*) for 1,14,15, 16,30,88,89,90,111,118,166,267,268,354,355,360,365,384, and 385 days.
Ff. 38r–38v. Table 22 of the *Makaranda*. But each entry is 1,0 diminished by the corresponding entry in table 22.
Ff. 38v–39v. Table 24 of the *Makaranda*. But each entry is 1,0 diminished by the corresponding entry in table 24.
Ff. 40r–41v. Blank.
Ff. 42–55 (4675). The *Śīghrabodha* of Kāśīnātha.

4878 (Smith Indic 21) ff. 1–2. On f. 1r a recent hand has written: śloka 100.
Ff. 1r–2v. Table of the Anonymous of 1714.

4879 (Smith Indic 173) 5 ff. On f. 5v, in the right-hand margin, is written in European numerals: 6415.
F. 1r. Blank.
Ff. 1v–5v. The *Makarandavivaraṇa* of Divākara. Incomplete.

4880 (U. Penn. 1838) 2 ff.
Ff. 1r–2v. The first 39½ verses of Divākara's *Makarandavivaraṇa*.

4881 (U. Penn. 1837) 20 ff. On f. 1r is written: atha makaraṃdasāraṇī likhyate śrīgurubhyo namaḥ //.
F. 1v. Table 1 of the *Makaranda* for Śaka 1688 to 1848 (A.D. 1766 to 1926).
F. 2r. Table 2 of the *Makaranda* for 1 to 16 years.
F. 2v. Table 3 of the *Makaranda* for 0 to 26 pakṣas.
Ff. 3r–3v. Table 4 of the *Makaranda* for 0 to 59 horizontal, and, in intervals of 6, for 0 to 54 vertical.
F. 4r. Table 5 of the *Makaranda* for Śaka 1664 to 1928 (A.D. 1742 to 2006).
F. 4r. Table 6 of the *Makaranda* for 1 to 24 years.

F. 4v. Table 7 of the *Makaranda* for 0 to 14 sidereal months.
Ff. 5r–5v. Table 8 of the *Makaranda* for 0 to 59 horizontal, and, in intervals of 6, for 0 to 54 vertical.
F. 6r. Table 10 of the *Makaranda* for 1 to 24 years.
F. 6r. Table 9 of the *Makaranda* for Śaka 1664 to 1904 (A.D. 1742 to 1982).
F. 6v. Table 11 of the *Makaranda* for 0 to 15.
Ff. 7r–7v. Table 12 of the *Makaranda* for 0 to 59 horizontal, and. in intervals of 6, for 0 to 54 vertical.
F. 8r. Table 13 of the *Makaranda* for Śaka 1688 to 1832 (A.D. 1766 to 1910).
F. 8r. Table 15 of the *Makaranda* for 1 to 12 zodiacal signs.
F. 8r. Table 14 of the *Makaranda* for 1 to 24 years.
F. 8r. Table 16 of the *Makaranda* for 1 to 27 nakṣatras.
F. 8v. Table 16a (of the *Makaranda?*) for Śaka 1680 to 1760 (A.D. 1758 to 1838), and for 1 to 20 years.
F. 8v. Table 16b (of the *Makaranda?*) for 1 to 12 zodiacal signs.
F. 9r. Table 17 of the *Makaranda* for Śaka 1685 to 1799 (A.D. 1763 to 1877), for 1 to 57 years, and for 1 to 26 pakṣas.
F. 9v. Table 18 of the *Makaranda*.
F. 10r. Table 19 of the *Makaranda*.
F. 10v. Table 20 of the *Makaranda*.
F. 11r. Table 21 of the *Makaranda*.
F. 11v. Table 22 of the *Makaranda*.
F. 12r. Table 23 of the *Makaranda*.
F. 12v. Table 24 of the *Makaranda*.
F. 13r. Table 25 of the *Makaranda*.
F. 13v. Table 26 of the *Makaranda*.
F. 14r. Table 27 of the *Makaranda* for 0° to 90° (combined in one table). It is added above this: śake 1541 yugagatavarṣāṇi 382720 guṇaka. Śaka 1541 is A.D. 1619, which is 4720 years from the beginning of Kaliyuga.
F. 14v. Table 28 of the *Makaranda* for 0° to 90°.
Ff. 15r–15v. Table 29 of the *Makaranda* for 0° to 180°. The parameters for the Greek-letter phenomena are not given.
Ff. 16r–16v. Table 30 of the *Makaranda* for 0° to 180°. The parameters for the Greek-letter phenomena are not given.
Ff. 17r–17v. Table 31 of the *Makaranda* for 0° to 180°. None of the additional parameters is given.
Ff. 18r–18v. Table 32 of the *Makaranda* for 0° to 180°. The parameters for the Greek-letter phenomena are not given.
Ff. 19r–19v. Table 33 of the *Makaranda* for 0° to 180°. The parameters for the Greek-letter phenomena are not given.
F. 20r. Table 36 (not filled in) (of the *Makaranda?*) for 1, 13, 14, 15, 16, and 17 days.
F. 20v. Blank.

4882 (U. Penn. 1788) ff. 1–25a and 25b–38. On f. 1r is written, in the lower margin: makaraṃdasāriṇī patra 39. Written in Argalāpura (*cf.* ff. 35r and 35v).
F. 1r. Table 1 of the *Makaranda* for Śaka 1624 to 1976 (A.D. 1702 to 2054). Above the table, beside the verse, is written: kāśyāṃ deśāṃtaraṃ sekaḥ dhanaṃ 47 vārādau vikramādityarājyaḥ syāt paṃcatriṃśottaraṃ śataṃ pātayitvā bhavechākaś caitraśuklād iti kramāt.
F. 1r. Table 2 of the *Makaranda* for 1 to 16 years.
F. 1v. Table 3 of the *Makaranda* for 0 to 27 pakṣas.
Ff. 2r–6v. Table 4 of the *Makaranda* for 0 to 59 horizontal, and 0 to 59 vertical.
F. 7r. Table 5 of the *Makaranda* for Śaka 1400 to 1760 (A.D. 1478 to 1838).
F. 7r. Table 6 of the *Makaranda* for 1 to 24 years.
F. 7v. Table 7 of the *Makaranda* for 0 to 14 sidereal months.
Ff. 8r–12v. Table 8 of the *Makaranda* for 0 to 59 horizontal, and 0 to 59 vertical.
F. 13r. Table 9 of the *Makaranda* for Śaka 1592 to 1880 (A.D. 1670 to 1958).
F. 13r. Table 10 of the *Makaranda* for 1 to 24 years.
F. 13v. Table 11 of the *Makaranda* for 0 to 14.
Ff. 14r–18v. Table 12 of the *Makaranda* for 0 to 59 horizontal, and 0 to 59 vertical.
F. 19r. Table 13 of the *Makaranda* for Śaka 1400 to 1976 (A.D. 1478 to 2054).
F. 19r. Table 14 of the Makaranda for 1 to 24 years.
F. 19r. Table 15 of the *Makaranda* for 1 to 12 zodiacal signs.
F. 19r. Table 16 of the *Makaranda* for 1 to 27 nakṣatras.
F. 19v. Table 17 of the *Makaranda*.
F. 20r. Table 18 of the *Makaranda*.
F. 20v. Table 19 of the *Makaranda*.
F. 21r. Table 20 of the *Makaranda*.
F. 21v. Table 21 of the *Makaranda*.
F. 22r. Table 22 of the *Makaranda*.
F. 22v. Table 23 of the *Makaranda*.
F. 23r. Table 24 of the *Makaranda*.
F. 23v. Table 25 of the *Makaranda*.
F. 24r. Table 26 of the *Makaranda*.
F. 24v. Table 27 of the *Makaranda* for 0° to 90°.
F. 25ar. Table 28 of the *Makaranda* for 0° to 90°.
F. 24av. Blank.
Ff. 25br–26r. Table 29 of the *Makaranda* for 0° to 180°.
Ff. 26v–27v. Table 30 of the *Makaranda* for 0° to 180°.
Ff. 28r–29r. Table 31 of the *Makaranda* for 0° to 180°.
Ff. 29v–30v. Table 32 of the *Makaranda* for 0° to 180°.
Ff. 31r–32r. Table 33 of the *Makaranda* for 0° to 180°.
Ff. 32r–33r. Table 34 (of the *Makaranda?*) for 1 to 108 nakṣatracaraṇas.
Ff. 33r–33v. Table 35 (of the *Makaranda?*) for 32, 31, 30, 29, 8 and 7 days.
Ff. 33v–34r. Table 36 (of the *Makaranda?*) for 1, 13, 14, 15, 16, and 17 days, and for 354, 355, 384, and 385 days.
F. 34r. Table 37 (of the *Makaranda?*).
F. 34v. Blank.
F. 35r Table 38 (of the *Makaranda?*) for Argalāpura.

The year is sidereal. The maximum length of daylight is 34;16 ghaṭikās at Cancer 0°.
F. 35r. Table 39 (of the *Makaranda?*).
F. 35v. Table 40 (of the *Makaranda?*) for Śaka 1700 to 1790 (A.D. 1778 to 1868), and for 1 to 30 years.
F. 35v. Table (not filled in) for Śaka 1850 to 2100 (A.D. 1928 to 2178) at intervals of 25 years. Its title is: atha agast⟨y⟩a udayāstasāraṇī // vṛṣe ⟨'⟩rke XX 1,0.
F. 35v. Table (not filled in) whose title is: siṃhe ⟨'⟩rke gatāṃśāt 4;26,20 udayaṃ munī dakṣiṇe argalāpure.
Ff. 36r–36v. Table 41 (of the *Makaranda?*).
F. 37r. Text on eclipses.
F. 37r. Table 42 (of the *Makaranda?*) for 1 to 36 viṃṣopakas.
F. 37v. Table 43 (of the *Makaranda?*) for 1 to 17 natas.
F. 37v. Table 44 (of the *Makaranda?*) for 1 to 36 daśakas.
F. 38r. Table 45 (of the *Makaranda?*) for 0 to 35 daśakas.
F. 38v. Table 46 (of the *Makaranda?*) for daśakas 34, 35, 0, 1, 2, 16, 17, 18, 19, and 20.
F. 38v. Table 47 (of the *Makaranda?*) for 1 to 24.

4883 (Smith Indic 34) ff. 9–11.

Ff. 9r–11r. Table 4 of the *Candrārkī* for 60° to 359°.
F. 11v. Blank.

4884 (Smith Indic 100 ff. 21r–22v). See 4904.

4886 (Smith Indic 98) 1 f. On f. 1r is written: śrīgaṇeśāya namaḥ // atha māsakoṣṭaka likhyate // saṃvat 1861, śāke 1727 [1726] vaiśākha śu di 14 ravau. This date corresponds to Sunday, 22 May 1804.
Ff. 1r–1v. A māsapraveśapattra. The maximum is 3;37,21 days (= 31;37,21 days) at Cancer 3°, the minimum 1;19,32 days (= 29;19,32 days) at Capricorn 1° to 3°. The names of the signs in the vertical column of arguments need to be lowered one place. The table is a duplicate of 4887.

4887 (Smith Indic 86) 1 f. On f. 1r, at the bottom, is written in European numerals: 6576.
Ff. 1r–1v. A māsapraveśapattra. The maximum is 3;37,21 days (= 31;37,21 days) at Gemini 3°, the minimum 1;19,32 days (= 29;19,32 days) at Sagittarius 1° to 3°.

4889 (Smith Indic 113) 2 ff. On f. 1r, in the bottom right-hand corner, is written in European numerals: 6571. On f. 2v is written: śloka 40.
Ff. 1r–2r. A dinamānapattra for Aries 0° to Pisces 29°. Half-lengths of daylight are tabulated. The maximum length of daylight is 33;20 ghaṭikās at Gemini 0°.
F. 2v. An explanatory text beginning:
meṣādike sāyanabhāgasūrye . . .

4892 (Smith Indic 138 ff. 15r–23v). See 4855.

4894 (Smith Indic 23) 2 ff.: A and B. On A f. 1r, at the top, is written: śloka 75.
A f. 1r. A dinamānapattra. The maximum length of daylight is 33;34 ghaṭikās at Gemini 11°. This indicates, at a precessional rate of 1° every 66 years, a date in the middle of the eighteenth century.
F. 1v. A table of the lords of the Decans (according to the Greek method).
F. 1v. A table of the diurnal and nocturnal lords of the triplicities.
F. 1v. A table said to be of the lords of the Decans, but really of the diurnal, nocturnal, and common lords of the triplicities.
F. 1v. A table of intercalary months for Śaka 1495 to 1727 (A.D. 1573 to 1805); preceding it is the following verse:

śākaḥ ṣaḍagnitithibhī 1536 rahito 'ṅkacandrair 19
bhaktaḥ śivābhragajabhūpaśarāniladigbhiḥ /
śeṣaṃ mitaṃ yad iha cet madhuto 'dhimāsā
jñeyās trayodaśamitaṃ yadi vā nabhaḥ syāt // 1 //

cai 11 / vai 0 / jye 8 / ā 16 / śrā 5 / bhā 2 / ā 10 / śrā 13

1495	Śrāvaṇa	1593	Vaiśākha
1498	Jyeṣṭha	1595	Bhādrapada
1500	Kārttika	1598	Śrāvaṇa
1503	Bhādrapada	1601	Jyeṣṭha
1506	Āṣāḍha	1603	Āśvina
1509	Vaiśākha	*1606*	Śrāvaṇa
1511	Śrāvaṇa	1609	Āśvina
1514	Bhādrapada	1612	Vaiśākha
1517	Jyeṣṭha	1614	Bhādrapada
1519	Āṣāḍha	1617	Śrāvaṇa
1522	Bhādrapada	1620	Jyeṣṭha
1525	Āṣāḍha	1622	Āśvina
1528	Caitra	*1625*	Śrāvaṇa
1530	Bhādrapada	1628	Āśvina
1533	Śrāvaṇa	1631	Vaiśākha
1536	Jyeṣṭha	1633	Bhādrapada
1538	Āśvina	1636	Śrāvaṇa
1541	Śrāvaṇa	1639	Jyeṣṭha
1544	Āṣāḍha	1641	Āśvina
1547	Caitra	*1644*	Śrāvaṇa
1549	Śrāvaṇa	1647	Āṣāḍha
1552	Āṣāḍha	1650	Vaiśākha
1555	Vaiśākha	1652	Bhādrapada
1557	Bhādrapada	1655	Śrāvaṇa
1560	Śrāvaṇa	1658	Jyeṣṭha
1563	Jyeṣṭha	1660	Āśvina
1566	Caitra	*1663*	Śrāvaṇa
1568	Śrāvaṇa (1569 in ms.)	1666	Āṣāḍha
1571	Āṣāḍha	1669	Caitra
1574	Vaiśākha	1671	Bhādrapada
1576	Bhādrapada	1674	Śrāvaṇa
1579	Śrāvaṇa	1677	Jyeṣṭha
1582	Jyeṣṭha	1680	Āśvina
1585	Caitra	*1682*	Śrāvaṇa
1587	Śrāvaṇa	1685	Āṣāḍha
1590	Āṣāḍha	1688	Vaiśākha

1690 Bhādrapada	1709 Bhādrapada
1693 Śrāvaṇa	1712 Śrāvaṇa
1696 Jyeṣṭha	1715 Jyeṣṭha
1699 Āśvina	1718 Āśvina
1701 Śrāvaṇa	*1720* Śrāvaṇa
1704 Āṣāḍha	*1727* Āṣāḍha
1707 Caitra	

B ff. 1r–1v. A dinamānapattra; discussed by O. H. Schmidt in *Isis* 35, 1944, 203–211, with a reproduction of the manuscript on 208. The maximum length of daylight is 33;34 ghaṭikās at Gemini 11°.

4895 (Smith Indic 40) 4 ff. On f. 1r, in the upper left-hand corner, is written: śloka 100.
Ff. 1r.–4v. Table 1 of the *Candrārkī* for 0 to 365 days.

4896 (U. Penn. 705) 11 ff. and 1 blank f. On f. 1r is written: śrīgaṇeśāya namaḥ ravyādigrahāṇāṃ madhyamasāraṇīyam // aharggaṇa ṣaṣṭibhakta.
Ff. 1r.–1v. Table 1 of the Anonymous of 1520 for 1 to 60 days, and for 1,0 to 1,10,0 days.
F. 1v. Table 2 of the Anonymous of 1520 for periods 20 to 30; dhruva given.
Ff. 2r.–2v. Table 3 of the Anonymous of 1520 for 1 to 60 days, and for 1,0 to 1,10,0 days.
F. 2v. Table 4 of the Anonymous of 1520 for periods 21 to 31; dhruva given.
Ff. 3r–3v. Table 5 of the Anonymous of 1520 for 1 to 60 days, and for 1,0, to 1,10,0 days.
F. 3v. Table 6 of the Anonymous of 1520 for periods 22 to 31; dhruva given.
Ff. 4r–4v. Table 7 of the Anonymous of 1520 for 1 to 60 days, and for 1,0 to 1,10,0 days.
F. 4v. Table 8 of the Anonymous of 1520 for periods 23 to 31; dhruva given.
Ff. 5r–5v. Table 9 of the Anonymous of 1520 for 1 to 60 days, and for 1,0 to 1,10,0 days.
Ff. 6r–6v. Table 11 of the Anonymous of 1520 for 1 to 60 days, and for 1,0 to 1,10,0 days.
F. 6v. Table 12 of the Anonymous of 1520 for periods 21 to 31; dhruva given.
Ff. 7r–7v. Table 13 of the Anonymous of 1520 for 1 to 60 days, and for 1,0 to 1,10,0 days.
F. 7v. Table 14 of the Anonymous of 1520 for periods 21 to 31; dhruva given.
Ff. 8r–8v. Table 15 of the Anonymous of 1520 for 1 to 60 days, and for 1,0 to 1,10,0 days.
F. 8v. Table 16 of the Anonymous of 1520 for period 28 (1728–1738 A.D.).
Ff. 9r–9v. Table 17 of the Anonymous of 1520 for 1 to 60 days, and for 1,0 to 1,10,0 days.
F. 9v. Table 18 of the Anonymous of 1520 for periods 22 to 31; dhruva given.
Ff. 10r–10v. Table 19 of the Anonymous of 1520 for 0° to 90°. Longitude of solar apogee given as Gemini 18°.
Ff. 11r.–11v. Table 20 of the Anonymous of 1520 for 0° to 26°. Entries in column 4 (increment to mean daily motion) are given to 3 sexagesimal places; maximum increment is 1;8,15,14°.

4903 (Smith Indic 96) 2 ff.
Ff. 1r–2v. Bhāvapattra from Aries 0° to Pisces 29°. The ascendent when Aries 0° is culminating is Cancer 6;4°; this places the latitude assumed in computing the table approximately at Ujjain.

4904, 4805, 4916, 4821, 4822, 4884 (Smith Indic 100) ff. 3–22. On F. 8r in the upper margin is written in European numerals: 6592.
Ff. 3r–7v (4904). Table 1 of the Anonymous of 1704 for 0 to 365 days.
Ff. 8r–9r (4805). Table 2 of the Anonymous of 1704 for 0 to 88 years.
Ff. 9v–11v (4916). Table 3 of the Anonymous of 1704 for 0 to 121 years. For the year 32 it notes Jye⟨ṣṭha⟩ as the intercalary month, and as the bīja of the Moon it gives − 15;35,24,3°.
Ff. 12r–12v (4821). Table 4 of the Anonymous of 1704 for 0 to 60 ghaṭikās.
F. 13r (4822). Table 5 of the Anonymous of 1704 for 0 to 42 years.
Ff. 13v–20v (4822). Table 6 of the Anonymous of 1704 for 0 to 365 years.
Ff. 21r–22v (4884). Table 7 of the Anonymous of 1704 for 0° to 90°.

4905 (U. Penn. 1847) 23 ff. On f. 1r. is written: atha laghucintāmaṇisāraṇeḥ prārambho ⟨')yam //; on f. 23v: iti laghucintāmaṇisāraṇeḥ samāptiḥ.
Ff. 1v–10r. Table 1 of the *Tithicintāmaṇi* for 0 to 400 tithis.
Ff. 10r–15r. Table 2 of the *Tithicintāmaṇi* for 0 to 390 nakṣatras.
Ff. 15r–23r. Table 3 of the *Tithicintāmaṇi* for 0 to 419 yogas.

4906 (U. Penn. 1891) 15 ff. On f. 1r is written: 2 atha
40
laghucintāmaṇisāriṇī prāraṃbhaḥ // 15 (*sic*); on f. 15v: śake 1683 vṛṣanāmasaṃvatsare kārta 3 kavaya 5 samāptaṃ / bhīmārpaṇam astu /. The date corresponds to 30 October 1761.
Ff. 1v–5v. Table 1 of the *Tithicintāmaṇi* for 0 to 400 tithis.
Ff. 6r–10r. Table 2 of the *Tithicintāmaṇi* for 0 to 390 nakṣatras.
Ff. 10v–15r. Table 3 of the *Tithicintāmaṇi* for 0 to 424 yogas.
F. 15r. Table 5 of the *Tithicintāmaṇi* for 1 to 12 zodiacal signs.
F. 15r. Table 4 of the *Tithicintāmaṇi* for 1 to 27 nakṣatras.
Ff. 15r–15v. Table 6 of the *Tithicintāmaṇi* for 27 nakṣatras.

4907 (U. Penn. 1848) 23 ff. This manuscript was written by the same scribe that wrote 4905. On f. 1r is written: atha laghutithicintāmaṇisāriṇer ārambhaḥ, and in pencil by a different hand: he pustaka dāmodara śāstri sahasrabuddhe.
Ff. 1v–9v. Table 1 of the *Tithicintāmaṇi* for 0 to 400 tithis.
Ff. 10r–15r. Table 2 of the *Tithicintāmaṇi* for 0 to 390 nakṣatras.
Ff. 15v–23v. Table 3 of the *Tithicintāmaṇi* for 0 to 400 yogas.

4916 (Smith Indic 100 ff. 9v–11v). See 4904.

4917 (Smith Indic 25) 1 f.
F. 1r. Table 1 of the Anonymous II of 1578 for Śaka 1501 to 1660 (A.D. 1579 to 1738).
F. 1v. Table 2 of the Anonymous II of 1578 for Śaka 1500 to 1666 (A.D. 1578 to 1744).

4923 (Smith Indic 190 ff. 3v–6v). See 4717.

4924 (Smith Indic 52) 2 ff. On f. 1r is written: śloka 600.
Ff. 1r–2v. Table 1 of the Anonymous of 1704 for 0 to 365 days.

4937 (Smith Indic 75) 1 f. On the verso is written: śloka 200, and, in European numerals: 6567.
F. 1r. A large table consisting of 81 squares (9 by 9); each of these large squares is divided into 81 small squares (9 by 9), each of which contains a number having two sexagesimal places. The entries within any large square form a consistent pattern, but there is no apparent relationship between the large squares. The purpose of the table is not known.

4942 (Smith Indic 119) ff. 2–18. On f. 2r, at the top, is written: ślo 300. On f.17v, at the bottom, is written: iti śrīsiddhaphali tithisāraṇī samāptā.
F. 2r. Verses 18c to 22 of an unidentified text. The colophon is: iti siddhaphali samāptā.
F. 2r. Table 1 of the Anonymous of 1594. The title is: śuddhyādīnāṃ guṇakāḥ brahma⟨pakṣe?⟩.
F. 2r. Table 2 of the Anonymous of 1594. The title is: śuddhyādīnāṃ kṣepakāḥ brahmapa⟨kṣe?⟩.
F. 2r. Table 3 of the Anonymous of 1594 for 1 to 12 zodiacal signs.
F. 2v. Table 4 of the Anonymous of 1594 for 1 to 27 nakṣatras.
F. 2v. Table 5 of the Anonymous of 1594.
Ff. 3r–7v. Table 6 of the Anonymous of 1594 for 0 to 27 horizontal, and for 0 to 59 vertical.
Ff. 8r–12v. Table 7 of the Anonymous of 1594 for 0 to 27 horizontal, and for 0 to 59 vertical.
Ff. 13r–17v. Table 8 of the Anonymous of 1594 for 0 to 29 horizontal, and for 0 to 59 vertical.
F.18r. Table 1 of the Anonymous of 1638 for 1 to 13 synodic months.
F. 18r. Table 2 of the Anonymous of 1638 for 1 to 14 sidereal months.
F. 18r. Table 3 of the Anonymous of 1638 for 1 to 13 synodic months.
F. 18r. Table 4 of the Anonymous of 1638 for 1 to 14 sidereal months.
F. 18r. Table 5 of the Anonymous of 1638 for Śaka 1560 to 1690 (A.D. 1638 to 1768).
F. 18v. Table 6 of the Anonymous of 1638 for 1 to 15.
F. 18v. Table 7 of the Anonymous of 1638 for 1 to 15.
F. 18v. Table 8 of the Anonymous of 1638 for 1 to 27 nakṣatras.
F. 18v. Table 9 of the Anonymous of 1638 for 1 to 12 zodiacal signs.

4943 (Smith Indic 138 f. 28). See 4855.

4944 (Smith Indic 172) 151 ff. On f. 1r, in the upper margin, is written: śloka 3000. On f. 150v, in the lower margin, is written: saṃvat 1814 nā varṣe vaiśāṣa su da 13 daṃne śanīvāsare paṃsāgarī pītābarajīsuta jevaṃtabhrātā prasotamabhrātā amarasī laṣate. This date corresponds to Saturday, 30 April 1757. On f. 151v, at the bottom, is written: śākamāthī 1626 (A.D. 1704).
Ff. 1r–30v. Table 9 of the *Mahādevī* (N = 0 to 59).
Ff. 31r–60v. Table 10 of the *Mahādevī* (N = 0 to 59).
Ff. 61r–90v. Table 11 of the *Mahādevī* (N = 0 to 59).
Ff. 91r–120v. Table 12 of the *Mahādevī* (N = 0 to 59).
Ff. 121r–150v. Table 13 of the *Mahādevī* (N = 0 to 59).
F. 151r. Table 20 of the *Jagadbhūṣaṇa* for 1 to 27 avadhis.
Ff. 151r–151v. Table 19 of the *Jagadbhūṣaṇa* for 0 to 92 years.

4945 (Smith Indic 194) 18 ff. On f.1v, in the upper left-hand corner, is written: śloka 500.
F. 1r. Blank.
Ff. 1v–6v. Table 1 of the *Tithicintāmaṇi* for 0 to 376 tithis.
F. 6v. A dinamānapattra listing not only the lengths of daylight for the days when the Sun is at the beginning of each zodiacal sign, but also cālakas measured in palas, a column headed: vārapravṛ⟨t⟩e palāni, and cālakas to that column. The greatest length of daylight occurs at Gemini 0°; it is 33;6 ghaṭikās. But clearly greater lengths must occur when the Sun is in Taurus.
Ff. 7r–12r. Table 2 of the *Tithicintāmaṇi* for 0 to 390 nakṣatras.
F. 12r. Table 5 of the Anonymous of 1594.
F. 12v. Table 16c (of the *Makaranda*).
F. 12v. Table of yearly parameters for the Lord of the Year, Epact, tithi, nakṣatra, yoga, tithimadhya, nakṣatramadhya, yogamadhya, tithispaṣṭa, nakṣatraspaṣṭa, and yogaspaṣṭa. *Cf.* table 8 (of the *Tithicintāmaṇi*).

Ff. 13r–18v. Table 3 of the *Tithicintāmaṇi* for 0 to 391 yogas.
F. 18v. Table 5 of the *Tithicintāmaṇi*.

4946 (Smith Indic M.B.) I 1 f.
F. 1r. Several verses, of which the first is:

śrīmanmaṅgalamūrtimārtrisamanaṃ natvā viditvā
 tataḥ
 śabdabrahmamanoramaṃ sugaṇako dānādhirā-
 mātmanaḥ /
yad granthādhyayane vineya nivahopācāryavaryām
 agāt
 so 'haṃ sūryakavir vilomaracanakāvyaṃ karomy
 adbhutam //1//

This has been identified as the beginning of Sūryakavi's *Rāmakṛṣṇavilomakāvya* (published in *Kāvyamālā*) by Prof. V. Raghavan of Madras.
F. 1v. Blank.

4946 (Smith Indic M.B.) II 4 ff.
Ff. 1–2. 9, 32–45 and 10, 1–17 of Varāhamihira's *Bṛhatsaṃhitā*. The manuscript begins: . . . rādhā jyeṣṭhāyāṃ kṣannamukhyasantāpaḥ.
F. 3. This f. contains (after salutations to Gaṇeśa, Sarasvatī, and Sūrya) only the verse:

jayati jagataḥ prasūtir viśvātmā sahajabhūṣaṇaṃ
 nabhasaḥ /
kratakanakasadṛśadaśaśatamayūkhamālārcitaḥ
 savitā // 1 //.

This is *Bṛhatsaṃhitā* 1, 1.
F. 4. This folio contains only:

gājīyaṃ śikhicāraṃ pārāśaram asitadevalakṛtaṃ ca /
anyāṃś ca bahūn dṛṣṭvā kriyate 'yam anākulaś
 cāraḥ // 1 //
darśanam astamayaṃ vā na gaṇitavidhir vāsaya
 śakyate jñātum /
divyāntarikṣabhaumās trividhāḥ syuḥ ketavo
 yasmāt // 2 //.

This is *Bṛhatsaṃhitā* 11, 1–2.

4946 (Smith Indic M.B.) III 1 f. On f. 1r, in the right-hand margin, is written in European numerals: 6569.
F. 1r. Table of lagnamānāni for Laṅkā (the Equator), Ujena (Ujjain), Gujarāti (Gujarat), Dakṣiṇa (the Deccan), Mulatāna (Multan), Bīkānera (Bikaner), Yojapura (Jodhpur), Nāgora (Nagaur), Rohitha (Rohtak), Jālora (Jalor), Rājapura (Rajpura), Bhṛguṭa (?), Ṣaṃbhāta (Sambhar), Marahadvala (?), Kanoja (Kanauj), Bhaṭevara (?), Kaśī (Benares), Ahammadā (Ahmadabad), Kāśmīra (Kashmir), Hiṃśāra (Hissar), Lāhora (Lahore), Āgarā (Agra), Kurukṣetra (Kurukṣetra), Dalli (Delhi), Jāladhara (Jullundur), Bahaṇi (?), Aṇaha (?), Navīnagara (Navanagar), Udayapura (Udaipur), Kānora (?), and, in the margin, Junāgaḍha (Junagarh). The lagnamānāni (rising-time in palas) for Laṅkā are:

Aries	278	Pisces
Taurus	299	Aquarius
Gemini	323	Capricorn
Cancer	323	Sagittarius
Leo	299	Scorpio
Virgo	278	Libra

F. 1v. Outline of a table, which is not filled in.

4946 (Smith Indic M.B.) IV. 1 f.
Ff. 1r–1v. A dinamānapattra for 0° to 29° of each of the 1 to 12 zodiacal signs. The maximum length of daylight is 33;34 ghaṭikās at Gemini 9°. Above the table is the remark: aināṃśā (*sic!*) 21. At 1° every 66 years, this amount of precession indicates a date toward the end of the nineteenth century.

4946 (Smith Indic M.B.) V 1 f.
F. 1r. A dinamānapattra for 0° to 29° of each of the 1 to 12 zodiacal signs. The maximum length of daylight is 33;34 ghaṭikās at Gemini 10°.
F. 1v. Blank.

4946 (Smith Indic M.B.) VI 1 f.
F. 1r. A dinamānapattra for 0° to 29° of each of the 1 to 12 zodiacal signs. The maximum length of daylight is 33;34 ghaṭikās at Gemini 11°.
F. 1v. A ⟨s⟩āyanalagnapattra for 0° to 29° of each of the 1 to 12 zodiacal signs. The text states: ayanāṃśa 20, which indicates a date in the early nineteenth century. The table records the accumulated rising-times in ghaṭikās for each degree of the zodiac as seen from some unknown latitude. The vernal equinox is at Pisces 10°.

4946 (Smith Indic M.B.) VII 1 f.
Ff. 1r–1v. A ⟨sāyana⟩ lagnapattra for 0° to 29° of each of the 1 to 12 zodiacal signs. The parameters given are: saurāṣṭre 'kṣaprabhā (noon equinoctial shadow in Saurāṣṭra): 5;0 (digits); ayanāṃśa (precession): 21°; akṣāṃśāḥ (terrestial latitude): 22;35,39°; akṣakarṇa (hypotenuse of the noon equinoctial shadow): 13 (digits); carakhaṇḍāni (?): 5;40,17. The vernal equinox is at Pisces 9°. Also given are the rising-times of the zodiacal signs measured in palas:

Aries	228 (= 22;48°)	Pisces
Taurus	259 (= 25;54°)	Aquarius
Gemini	306 (= 30;36°)	Capricorn
Cancer	340 (= 34;0°)	Sagittarius
Leo	339 (= 33;54°)	Scorpio
Virgo	328 (= 32;48°)	Libra

4946 (Smith Indic M.B.) VIII 1 f.
Ff. 1r.–1v. A sāyanalagnapattra for 0° to 29° of each of the 1 to 12 zodiacal signs. Precession here equals 22°,

which indicates a date in the middle of the twentieth century! The vernal equinox is at Pisces 8°.

4946 (Smith Indic M.B.) IX f. 1.
F. 1r. A double-entry astrological table for 1 to 28 (nakṣatras) horizontal, and for 1 to 7 planets, the yoga, and the auspiciousness vertical. The entries opposite the first 7 vertical arguments are names of nakṣatras, opposite the 8th names of yogas, and opposite the 9th an indication of whether it is auspicious or not.
F. 1v. Blank.

4946 (Smith Indic M.B.) X 1 f.
F. 1r. A sāyanalagnapattra for 0° to 29° of each of the 1 to 12 zodiacal signs for Laṅkā (the equator). The vernal equinox is at Pisces 10°.
F. 1v. Blank.

4946 (Smith Indic M.B.) XI 1 f. On f. 1v, in the upper right-hand margin, is written: śloka 120.
F. 1r. Table 1 of the Anonymous of 1790 for Śaka 1712 to 1808 (A.D. 1790 to 1896).
F. 1r. Table 2 of the Anonymous of 1790 for 1 to 24 years.
F. 1r. Table 3 of the Anonymous of 1790 for 1 to 13 synodic months.
F. 1r. Table 4 of the Anonymous of 1790 for 1 to 14 sidereal months.
F. 1v. Table 5 of the Anonymous of 1790 for 1 to 15 periods.

4946 (Smith Indic M.B.) XII 1 f. On f. 1v, at the bottom, is written: saṃvat 1899 nā śāke 1764 pra° āṣāḍha vi di 12 budho likhitaṃ puṣkarṇājnā° josī śrīpa mukadajīsuta. This date corresponds to Wednesday, 3 August 1842.
Ff. 1r–1v. Table of the increment to or decrease from the mean daily motion of the Sun for 0 to 365 days.

4946 (Smith Indic M.B.) XIII 1 f. On f. 1r. in the right-hand margin, is written in European numerals: 6596.
Ff. 1r–1v. A double-entry table for 1 to 27 nakṣatras horizontal (the title at the top of f. 1r is: ravikoṣṭhaka), and for 1 to 27 nakṣatras vertical (the title in the left-hand margin of f. 1r is: rāhukoṣṭhaka). The purpose of this table is not understood.

4946 (Smith Indic M.B.) XIV ff. 3–11. On f. 3r is written: śloka 229.
F. 3r. Table 1 of the *Karaṇakesari* for 1 to 130 periods of 130 years.
F. 3v. Table 2 of the *Karaṇakesari* for 0 to 130 years.
F. 4r. Table 3 of the *Karaṇakesari* for 1 to 27 avadhis.
F. 4r. Table 4 of the *Karaṇakesari* for 1° to 16°.
F. 4r. Table 5 of the *Karaṇakesari* in 20 entries.
F. 4v. Table 6 of the *Karaṇakesari* for 0;52 to 1;7 days.
F. 4v. Table 7 of the *Karaṇakesari* for 1 to 12 zodiacal signs.
F. 4v. Table 8 of the *Karaṇakesari* for 1 to 21 digits.
F. 4v. Table 9 of the *Karaṇakesari* for 1 to 9 digits.
F. 4v. Table 10 of the *Karaṇakesari* for 1 to 27 avadhis.
F. 4v. Table 11 of the *Karaṇakesari* for 1 to 21 digits.
F. 5r. Table 12 of the *Karaṇakesari* for 1 to 21 digits.
F. 5r. Table 13 of the *Karaṇakesari* for 1 to 21 digits.
Ff. 5r–5v. Table 14 of the *Karaṇakesari* for 0° to 45°.
F. 5v. Table 15 of the *Karaṇakesari* for 9 parvas.
F. 6r. Table 16 of the *Karaṇakesari* for 0° to 90°.
F. 6v. Table 17 of the *Karaṇakesari* for 0 to 47 (degrees?).
F. 6v. Table 18 of the *Karaṇakesari* for 0 to 47 (degrees?).
F. 7r. Table 19 of the *Karaṇakesari* for 0° to 29° horizontal, 0 to 11 zodiacal signs vertical.
Ff. 7v–8r. Table 20 of the *Karaṇakesari* for 0° to 29° horizontal, 0 to 11 zodiacal signs vertical.
Ff. 8v–9v. Table 21 of the *Karaṇakesari* for 1 to 60 horizontal, 1 to 6 pairs of zodiacal signs vertical.
F. 9v. Table 22 of the *Karaṇakesari* for 0° to 91°.
F. 10r. Table 23 of the *Karaṇakesari* for 0° to 29° horizontal, 0 to 11 zodiacal signs vertical.
F. 10v. Table 24 of the *Karaṇakesari* for 0° to 29° horizontal, 0 to 11 zodiacal signs vertical.
F. 11r. Table 25 of the *Karaṇakesari* for 1 to 12 digits.
F. 11r. Table 26 (of the *Karaṇakesari?*) for 1 to 60.
F. 11v. Table 27 (of the *Karaṇakesari?*) for 1 to 28 nakṣatras.

4946 (Smith Indic M.B.) XV f. 5. On f. 5r, in the upper left-hand corner, is written in European numerals: 6583; in the upper margin: śloka 200.
F. 5r. Table of aspects of the three superior planets for 0° to 29° horizontal, and for various zodiacal signs vertical. Identical with 4946 LVI. *q.v.*
F. 5v. Table of the horās of the Sun and Moon.
F. 5v. Table of the degrees in the three dreṣkāṇas in a sign.
F. 5v. Table of the degrees in the seven saptāṃśas in a sign.
F. 5v. Table of the members of the four triplicities.
F. 5v. Table of the degrees in the nine navāṃśas in a sign.
F. 5v. Table of the degrees in the twelve dvādaśāṃśas in a sign.
F. 5v. Table of the degrees in each of the planetary fines in odd signs and in even signs.
F. 5v. Verses 57b–62d of an unidentified text.

4946 (Smith Indic M.B.) XVI f. 1.
Ff. 1r–1v. Table 1 of the Anonymous of 1598; 1 to 20 years for Mars, Mercury, Jupiter, Venus, and Saturn.

4946 (Smith Indic M.B.) XVII 1 f.
F. 1r. Table of the exaltations and dejections of the seven planets.
F. 1v. Table of triplicities.
F. 1v. Table of the six groupings of nakṣatras.

4946 (Smith Indic M.B.) XVIII f. 1.
F. 1r. Text on planetary aspects.
F. 1v. Table of mean length of life from the aṣṭakavarga.
F. 1v. Table of corrected length of life from the aṣṭakavarga.

4946 (Smith Indic M.B.) XIX f. 7.
F. 7r. Diagram of "rain-cycle" (vṛṣṭicakra).
F. 7r. Diagram of "turtle-cycle" (kūrmacakra).
F. 7v. Table of characteristics of the seven planets in the twelve zodiacal signs.

4946 (Smith Indic M.B.) XX 1 f. On f. 1r at the bottom is written: saṃvat 1835 śāke 1700 (A.D. 1778). tadā grahalāghave cakramukho 'bdapavārādi 1;13,30,33, 18,36 guṇaka abdapano vārādi 1;15,30,49,46,25 śākamāhemthī 1700 hīnakīje śeṣaguṇakagaṇie kṣepaka 1;13, 30,33,18,36 bhelie. . .
Ff. 1r–1v. Table 1 of the Anonymous of 1747.

4946 (Smith Indic M.B.) XXI ff. 2 and 8, and an unnumbered f.
Ff. 2r–2v. Table 1 of the Anonymous of 1656 for the following periods:
Mars: Śaka 1578 to 2018 (A.D. 1656 to 2096).
Mercury: Śaka 1578 to 1938 (A.D. 1656 to 2016).
Jupiter: Śaka 1578 to 1938 (A.D. 1656 to 2016).
Venus: Śaka 1578 to 1678 (A.D. 1656 to 1756).
Saturn: Śaka 1578 to 1678 (A.D. 1656 to 1756).
Ff. 8r–8v. Table 2 of the Anonymous of 1656 for Śaka 1601 to 1685 (A.D. 1679 to 1763).
Unnumbered f. Table 3 of the Anonymous of 1656.

4946 (Smith Indic M.B.) XXII 1 f. On f. 1r is written, in the lower right-hand corner, in European numerals: 6525.
F. 1r. Blank.
F. 1v. Table 10 of the *Mahādevī* (N = 56 to 57).

4946 (Smith Indic M.B.) XXIII f. 20.
F. 20r. Table of the degrees of each of the 27 nakṣatras.
F. 20r. Table of 1 to 60 equal parts of a nakṣatra (13; 20°).
F. 20v. Blank except for the title: "Four-in-one table of the Lord of the Year" (abdapacatuṣṭaikya).

4946 (Smith Indic M.B.) XXIV f. 19.
Ff. 19r–19v. Table 3 of the *Candrārkī* for 1 to 60 ghaṭikās.

4946 (Smith Indic M.B.) XXV 1 f. On f. 1r, in the upper left-hand corner, is written: śloka 132.
F. 1r. Blank.
F. 1v. Table 10 of the *Mahādevī* (N = 1).

4946 (Smith Indic M.B.) XXVI 1 f.
F. 1r. Table of right ascensions of the zodiacal signs measured in palas.
Ff. 1r–1v. Table of accumulated rising-times measured in ghaṭikās for every degree from Aries 0° to Pisces 29° (Aquarius 25°–29° and Pisces 25°–29° are omitted); arranged 0° to 29° horizontal, 0 to 11 zodiacal signs vertical. The vernal equinox is at Pisces 9°.

4946 (Smith Indic M.B.) XXVII 2 ff. On f. 2r, in lower right-hand corner, is written in European numerals: 6568.
Ff. 1r–1v. Table 21 of the *Karaṇakesarī* for 1 to 60 horizontal, and for 1 to 6 pairs of zodiacal signs vertical.
Ff. 2r–2v. A similar table for 1 to 60 horizontal, and for 1 to 6 pairs of zodiacal signs vertical, but calculated for a different latitude and for a tropical zodiac:

Aries, Pisces	9;16	Virgo, Libra
Taurus, Aquarius	9;58	Leo, Scorpio
Gemini, Capricorn	10;46	Cancer, Sagittarius

4946 (Smith Indic M.B.) XXVIII 1 f.
Ff. 1r–1v. Table of the mahādaśās of the planets.

4946 (Smith Indic M.B.) XXIX 1 f.
F. 1r. Blank.
F. 1v. Table 1 of the *Mahādevī;* the labdha section is complete, the śeṣa section contains only entries for years 0 to 9 and 30 to 32.

4946 (Smith Indic M.B.) XXX 1 f.
F. 1r. Table 1 of the *Mahādevī.*
F. 1v. Table 2 of the *Mahādevī.*

4946 (Smith Indic M.B.) XXXI 1 f.
F. 1r. Table of the Lords of the Year from Śaka 1660 to 1740 (A.D. 1738 to 1818), computed according to the *Mahādevī.*
F. 1v. Table of the Epacts from Śaka 1660 to 1740 (A.D. 1738 to 1818), computed according to the *Mahādevī. Cf.* 4946 LXXI.

4946 (Smith Indic M.B.) XXXII f. 4. On f. 4v, in the upper right-hand corner, is written: śloka 30; in the lower left-hand corner, in European numerals: 6577.
F. 4r. Table of increases to (or decreases from) the mean daily progress of the Sun for days 200 to 365.
F. 4r. Verses 14–16 of an unidentified text.
F. 4v. Table of the beginnings of the 27 nakṣatras.
F. 4v. Table of the beginnings of the 108 nakṣatrapādas.

4946 (Smith Indic M.B.) XXXIII 1 f.
Ff. 1r–1v. Table indicating the following information for the beginnings of 1 to 104 years: (a) week-day. (b)

fraction of a day. The year-length is 6,5;15,31,30 days. (c) nakṣatra. (d) tithi.

4946 (Smith Indic M.B.) XXXIV f. 138. On f. 138v is written: saṃvat 1899 nā śāke 1764 pra śubha āṣāḍha vidi 11 bhaume likhitaṃ puṣkarṇajñātīyaṃ avaṭaṃkapaṇiyā josī śrī 5 mukuṃdajīsuta doṣā svayaṃ paṭhanārthaṃ // grāma śrīmaṃḍanapure. The date corresponds to Tuesday, 2 August 1842.
F. 138r. Table 19 of the *Jagadbhūṣaṇa* for 0 to 92 years.
F. 138v. Table 20 of the *Jagadbhūṣaṇa* for 1 to 27 avadhis.

4946 (Smith Indic M.B.) XXXV 1 f.
Ff. 1r–1v. Table 9 of the Anonymous of 1704 (N = 30).

4946 (Smith Indic M.B.) XXXVI 1 f.
F. 1r. Table of the beginnings of the 27 nakṣatras.
F. 1r. Table 37 (of the *Makaranda?*).
F. 1v. Blank.

4946 (Smith Indic M.B.) XXXVII ff. 1–2. On f. 2v, in the upper left-hand corner, is written: śloka 144; in the lower right-hand corner, in European numerals: 6588.
F. 1r. Table of horās.
F. 1r. Table of drekāṇas.
F. 1r. Table of caturthāṃśas.
F. 1r. Table of pañcamāṃśas.
F. 1r. Table of ṣaṣṭhāṃśas.
F. 1v. Table of saptamāṃśas
F. 1v. Table of aṣṭamāṃśas.
F. 1v. Table of navamāṃśas.
F. 2r. Table of daśamāṃśas.
F. 2r. Table of ekādaśāṃśas.
F. 2v. Table of dvādaśāṃśas.
F. 2v. Table of triṃśāṃśas (fines).
F. 2v. Table of friendships.

4946 (Smith Indic M.B.) XXXVIII 1 f.
F. 1r. First 8½ verses of the *Candrārkī*.
F. 1v. Blank.

4946 (Smith Indic M.B.) XXXIX 1 f.
F. 1r. First 7 verses of the *Candrārkī*.
F. 1r. Table of parameters of the *Candrārkī*:

	Yearly motions	Epoch positions	Rāmabījas
Lord of the Year	1;15,31,17,17	1;24,57,12,30	0;0,4
Epact	11;3,53,22,40	22;18,34,10,0	0;1,15
Kendra	7;40,30,40,20	21;52,16,31,18	−0;3,45

F. 1v. Blank.

4946 (Smith Indic M.B.) XL ff. 15–22.
F. 15r. Table 2 of the Anonymous of 1704 for 31 to 88 years.
Ff. 15r–17r. Table 1 of the Anonymous of 1704 for 0 to 365 days.
Ff. 17v–18r. Table 3 of the Anonymous of 1704 for 0 to 121 years.
F. 18v. Table 5 of the Anonymous of 1704 for 0 to 42 years.
Ff. 18v–19r. Table 4 of the Anonymous of 1704 for 1 to 60 ghaṭikās.
Ff. 19r–22r. Table 6 of the Anonymous of 1704 for 0 to 365 days.
Ff. 22r–22v. Table 7 of the Anonymous of 1704 for 0° to 90°.

4946 (Smith Indic M.B.) XLI f. 54.
F. 54r. Verses 9c–14 of a work on ramalaśāstra (geomancy).
F. 54v. Blank.

4946 (Smith Indic M.B.) XLII 1 f.
Ff. 1r–1v. Excerpt from the *Ārambhasiddhi* of Udayaprabha.

4946 (Smith Indic M.B.) XLIII 3 ff.
Ff. 1r–3v. A *Praśnavinoda*.

4946 (Smith Indic M.B.) XLIV 1 f.
F. 1r. The first 6½ verses of an unidentified text.
F. 1v. Blank.

4946 (Smith Indic M.B.) XLV f. 1.
Ff. 1r–1v. Table of the accumulated rising-times of the degrees of the zodiac; arranged for 0° to 29° horizontal, and for 0 to 11 zodiacal signs vertical. The vernal equinox is at Pisces 8°.

4946 (Smith Indic M.B.) XLVI f. 6. On f. 6v, in the upper left-hand corner, is written in European numerals: 6579.
F. 6r. Table of "lords of the signs" (rāśīśadhruvacakra):

Sun	6	Mercury	17	Saturn	10
Moon	15	Jupiter	19		
Mars	8	Venus	21		

F. 6r. Table 5 of the Anonymous of 1594.
F. 6r. Table 14 of the Anonymous of 1741 for 1 to 27 nakṣatras.
F. 6v. Table 15 of the Anonymous of 1741 for 1 to 12 zodiacal signs.
F. 6v. Text reading: śāke hīna 508 śeṣaṃ sanna / saṃvate hīnaṃ 643 seṣaṃ śannaṃ / sane yutaṃ 643 jātaṃ āṣāḍhādī saṃvataṃ / sane yutaṃ 508 yātoyaṃ caitrādī śākāyam /. Sanna (Arabic سنة), then, equals 586 A.D.

4946 (Smith Indic M.B.) XLVII 1 f. On f. 1v, in the upper left-hand corner, is written in European numerals: 6582.
Ff. 1r–1v. Table of the daśās of the seven planets and Rāhu.

4946 (Smith Indic M.B.) XLVIII 1 f. On f. 1r, in the

upper left-hand corner, is written in European numerals: 6581.
F. 1r. Table giving the following attributes of each of the 28 nakṣatras: akṣaras, forms, deities, number of stars, animals, spheres (divine, human, or demonic), nāḍīs, and yuṃjās.
F. 1v. Table of the 28 nakṣatras, centering around Abhijit:

Aśvinī 3,9	Svāti 9,4
Bharaṇī 3,20	Viśākhā 9,14
Kṛttikā 4,4	Anurādhā 10,2
Rohiṇī 4,15	Jyeṣṭhā 10,15
Mṛgaśiras 5,0	Mūla 10,25
Ārdrā 5,6	Pūrvāṣāḍhā 11,15
Punarvasu 6,7	Uttarāṣāḍhā 11,24
Puṣya 6,14	Abhijit 0,0
Āśleṣā 6,25	Śravaṇa 0,21
Maghā 7,5	Dhaniṣṭhā 0,24
Pūrvaphālgunī 7,20	Śatabhiṣak 1,19
Uttaraphālgunī 7,28	Pūrvabhādrapadā 2,5
Hasta 8,9	Uttarabhādrapadā 2,15
Citra 8,20	Revatī 3,6

F. 1v. Melothesia for men and melothesia for women.
F. 1v. Table of relations between triplicities and the four elements.
F. 1v. Table of planetary friendships.

4946 (Smith Indic M.B.) XLIX f. 1. On f. 1v, in the upper right-hand corner, is written: śloka 40; in the lower right-hand corner, in European numerals: 6573.
Ff. 1r–1v. Table of accumulated rising-times of the degrees of the zodiac; arranged for 0° to 29° horizontal, and for 1 to 12 zodiacal signs vertical. The vernal equinox is at Pisces 8°.

4946 (Smith Indic M.B.) L f. 1. On f. 1v, in the upper left-hand corner, is written: śloka 50.
F. 1r. Table of week-days on which the Sun enters each of the 27 nakṣatras, and of a function which resembles the equation of daylight except that the minimum is too low:

Aśvinī	4;47,21	−0;4
Bharaṇī	4;28,27	+0;29
Kṛttikā	4;16,15	+0;51
Rohiṇī	4;10,5	+1;12
Mṛgaśiras	4;8,41	+1;25
Ārdrā	4;10,36	+1;17
Punarvasu	4;14,0	+1;7
Puṣya	4;17,2	+0;54
Āśleṣā	4;17,50	+0;38
Maghā	4;14,44	+0;27
Pūrvaphālgunī	4;6,15	+0;16
Uttaraphālgunī	3;51,30	+0;5
Hasta	3;29,50	−0;8
Citra	3;0,51	−0;22
Svāti	2;24,52	−0;37
Viśākhā	1;42,29	−1;0 (*sic* MS)
Anurādhā	0;54,25	−1;11
Jyeṣṭhā	0;1,47	−1;46
Mūla	6;5,56	−2;10
Pūrvāṣāḍhā	5;8,8	−2;26
Uttarāṣāḍhā	4;10,3	−2;27
Śravaṇa	3;12,56	−2;14
Dhaniṣṭhā	2;18,33	−2;7
Śatabhiṣak	1;27,37	−1;47
Pūrvabhādrapadā	0;41,53	−1;15
Uttarabhādrapadā	0;2,19	−0;42
Revatī	6;28,55	−0;7

F. 1r. Table of the beginnings of the nakṣatras and of their pādas.
F. 1v. Table of the week-days on which the sun enters each of the 12 zodiacal signs:

Aries	4;47,21	Libra	2;43,43
Taurus	0;44,15	Scorpio	4;36,54
Gemini	4;9,19	Sagittarius	6;5,56
Cancer	0;46,27	Capricorn	0;25,38
Leo	4;14,44	Aquarius	1;52,34
Virgo	0;16,50	Pisces	3;41,26

4946 (Smith Indic M.B.) LI f. 1.
Ff. 1r–1v. Table of the Lord of the Year for Śaka 1501 to 1690 (A.D. 1579 to 1768), calculated roughly according to the *Madādevī*. *Cf.* Anonymous II of 1578.

4946 (Smith Indic M.B.) LII 2 ff. On f. 1v. is written: śloka 129.
Ff. 1r-2v. Table of mean motions of the Epact, Lord of the Year, Moon, and Lunar Anomaly, using yearly motions and epoch positions identical with those of table 5 of the *Candrārkī*. But these tables are for 0 to 29 single years, and 1 to 30 groups of 30 collected years.

4946 (Smith Indic M.B.) LIII f. 1.
Ff. 1r–1v. Dinamānapattra for 0° to 29° horizontal, and for 1 to 12 zodiacal signs vertical. The longest day is 33;34,0 ghaṭikās when the Sun is at Gemini 9°.

4946 (Smith Indic M.B.) LIV f. 1.
F. 1r. Table of rising-times of the zodiacal signs measured in palas:

Aries	228	Pisces
Taurus	259	Aquarius
Gemini	306	Capricorn
Cancer	340	Sagittarius
Leo	339	Scorpio
Virgo	328	Libra

Ff. 1r–1v. Table of rising-times for 0° to 29° horizontal, and for 1 to 12 zodiacal signs vertical. The vernal equinox is at Pisces 9°. Identical with 4946 LV.

4946 (Smith Indic M.B.) LV 1 f.
Ff. 1r–1v. The same two tables as are found in 4946

LIV. The title is: kachasaurāṣṭre lagnapatraṃ ayanāṃśa 21.

4946 (Smith Indic M.B.) LVI f. 10.
Ff. 10r–10v. Table of the planetary aspects for 0° to 29° horizontal, and for the following zodiacal signs vertical:

a) Saturn: 1,2,8,9
b) Jupiter: 3,4,7,8
c) Mars: 2,3,7,6

F. 10r. Table of horās.
F. 10v. Table of drekāṇas.
F. 10v. Table of saptāṃśas.
F. 10v. Table of navāṃśas.
F. 10v. Table of dvādaśāṃśas.
F. 10v. Table of fines (triṃśāṃśas).
All these tables are found in 4946 XV.

4946 (Smith Indic M.B.) LVII ff. 1–2.
Ff. 1r–2v. Table of characteristics and attributes of the 12 zodiacal signs.

4946 (Smith Indic M.B.) LVIII ff. 1–5. On f. 1r, in the upper margin, is written: śloka 160.
F. 1r. Table 1 of the *Brahmatulyasāraṇī*. In this manuscript the longitudes for 1 to 30 periods of 20 "years" rather than of 1 to 60 are recorded.
F. 1v. Table 2 of the *Brahmatulyasāraṇī*.
F. 2r. Table 3 of the *Brahmatulyasāraṇī*.
F. 2v. Table 4 of the *Brahmatulyasāraṇī*.
F. 3r. Table 5 of the *Brahmatulyasāraṇī*.
F. 3v. Table 6 of the *Brahmatulyasāraṇī*.
F. 4r. Table 7 of the *Brahmatulyasāraṇī*.
F. 4v. Table 8 of the *Brahmatulyasāraṇī*.
F. 5r. Table 9 of the *Brahmatulyasāraṇī*.
F. 5v. Table 12 of the *Brahmatulyasāraṇī*.

4946 (Smith Indic M.B.) LIX 1 f.
Ff. 1r–1v. Table 11 of the *Mahādevī* (N = 10; copied twice).

4946 (Smith Indic M.B.) LX ff. 1–2.
Ff. 1r–2r. Table of true longitudes of the planets for Saṃ. 1849, Śaka 1714 (A.D. 1792), computed according to the Anonymous of 1704.
F. 2v. Unidentified astrological text on lagnaphala, etc.

4946 (Smith Indic M.B.) LXI f. 91. On f. 91v. is written: śloka 80; and: saṃvat 1860 nā jyeṣṭha śu da 2 śanau. This date corresponds to Saturday, 21 May 1803.
F. 91r. Table 13 of the *Mahādevī* (N = 58 to 59).
F. 91v. Blank.

4946 (Smith Indic M.B.) LXII f. 11. This is f. 11 of 4735.
Ff. 11r–11v. Table 17 of the *Brahmatulyasāraṇī*.

4946 (Smith Indic M.B.) LXIII. 4 ff.
F. 1r. Table 1 of the *Mahādevī*.
F. 1v. Blank.
F. 2r. Table 10 of the *Mahādevī* (N = 3).
F. 2v. Blank.
F. 3r. Table 10 of the *Mahādevī* (N = 13).
F. 3v. Blank.
F. 4r. Table 10 of the *Mahādevī* (N = 15).
F. 4v. Blank.

4946 (Smith Indic M.B.) LXIV. 1 f.
F. 1r. Table 3 of the *Mahādevī*.
F. 1v. Blank.

4946 (Smith Indic M.B.) LXV f. 26. On f. 26r, in the upper left-hand corner, is written: śloka 150.
Ff. 26r–26v. Table 10 of the *Mahādevī* (N = 18 to 19).

4946 (Smith Indic M.B.) LXVI ff. 1–2.
Ff. 1r–2v. Table 3 of the Anonymous of 1656.

4946 (Smith Indic M.B.) LXVII ff. 1–2. On f. 2v is written: śloka 250.
Ff. 1r–2r. Unidentified astrological text.

4946 (Smith Indic M.B.) LXVIII 1 f. On f. 1v, in the upper margin, is written: śloka 60.
F. 1r. Tables of mean motion for 1 to 9, 10 to 90, and 100 to 900 years for the Epact, Lord of the Year, Mars, Mercury, Jupiter, Venus, Saturn, and the Lunar Node, according to the *Mahādevī*.
F. 1v. Explanatory text.

4946 (Smith Indic M.B.) LXIX 1 f. On f. 1r, in the upper margin, is written: śloka 125.
F. 1r. Table of sixtieths of a nakṣatra.
F. 1r. Table of the beginnings of the 27 nakṣatras.
F. 1v. Unidentified text.

4946 (Smith Indic M.B.) LXX 2 ff. On f. 1r is written: śloka 62.
Ff. 1r–2v. Table of longitudes of the five planets for Saṃ. 1854, Śaka 1719 (A.D. 1797), computed according to the Anonymous of 1704.

4946 (Smith Indic M.B.) LXXI f. 2. On f. 2r, in the upper margin, is written: śloka 30.
F. 2r. Table of the Moon's mean motion from Śaka 1660 to 1743 (A.D. 1738 to 1821). The yearly mean motion is 2,12;46,40,33°; the epoch longitude 4,32;7,55°.
F. 2v. Table of the mean motion of the Moon's Anomaly from Śaka 1660 to 1731 (A.D. 1738 to 1809). The yearly mean motion is 1,32;6,8,52°; the epoch longitude 3,58;5,51°. *Cf.* 4946 XXXI

4946 (Smith Indic M.B.) LXXII. f. 4.
Ff. 4r–4v. Astrological text (commentary on a work on muhūrtas) concerning praśnas; quotes Nārada.

4946 (Smith Indic M.B.) LXXIII 1 f.
Ff. 1r–1v. Text on prastāvayoga.

4946 (Smith Indic M.B.) LXXIV. 1 f.

F. 1r. Table of the Lords of the Year for 1 to 108 years; the yearly parameter is 1;15,31,30,30 days.
F. 1v. Blank.
Cf. 4946 LXXXI

4946 (Smith Indic M.B.) LXXV f. 1.
Ff. 1r–1v. Tables of the lords of various subdivisions of the zodiacal signs.

4946 (Smith Indic M.B.) LXXVI f. 1.
Ff. 1r–1v. Table of the beginnings of the nakṣatrapādas of the 27 nakṣatras.

4946 (Smith Indic M.B.) LXXVII f. 1. On f. 1v is written: śloka 125.
F. 1r. Table 2 of the *Mahādevī*.
F. 1v. Table 1 of the *Mahādevī*.

4946 (Smith Indic M.B.) LXXVIII f. 3.
Ff. 3r–3v. Text on aspects and aṣṭakavarga from a work entitled *Muktāvalī*.

4946 (Smith Indic M.B.) LXXIX. f. 1.
Ff. 1r–1v. Table of accumulated rising-times of the degrees of the zodiac measured in ghaṭikās; arranged for 0° to 29° horizontal, and for 1 to 12 zodiacal signs vertical. The vernal equinox is at Aries 0°.

4946 (Smith Indic M.B.) LXXX f. 1.
Ff. 1r–1v. Table of the following functions at the beginning of 1 to 100 years: tithi, week-day, nakṣatra, and ascendent sign. The yearly parameter for the week-day (Lord of the Year) is 1;15,31,30 days.

4946 (Smith Indic M.B.) LXXXI f. 1.
Ff. 1r–1v. Duplicate of 4946 LXXX, but for 1 to 108 years.
Cf. 4946 LXXIV.

4946 (Smith Indic M.B.) LXXXII f. 1.
Ff. 1r–1v. Table of accumulated rising-times measured in ghaṭikās for the degrees of the zodiac; arranged for 0° to 29° horizontal, and for 1 to 12 zodiacal signs vertical. The vernal equinox is at Aries 0°.

4946 (Smith Indic M.B.) LXXXIII f. 1.
Ff. 1r–1v. Table of oblique ascensions for 0° to 29° horizontal, and for 1 to 12 zodiacal signs vertical. The vernal equinox is at Pisces 9°.

4946 (Smith Indic M.B.) LXXXIV f. 1. On f. 1v is written: śloka 300.
Ff. 1r–1v. Tables of aṣṭakavargas.

4946 (Smith Indic M.B.) LXXXV. 1 f. On f. 1v is written: śloka 100.
F. 1r. Blank.
F. 1v. Unidentified text.

4946 (Smith Indic M.B.) LXXXVI f. 1.
Ff. 1r–1v. Table 3 of the Anonymous of 1656.

4946 (Smith Indic M.B.) LXXXVII 1 f.
Ff. 1r–1v. Unidentified astrological text from *Jā(takā)-bha(raṇa?* of Ḍhuṇḍhirāja?).

4946 (Smith Indic M.B.) LXXXVIII 1 f.
F. 1r. Table 2 of the *Candrārkī* for 0 to 63 days.
F. 1v. Blank.

4946 (Smith Indic M.B.) LXXXIX f. 1.
Ff. 1r–1v. Table 1 of the *Candrārkī* for 0 to 103 days.

4946 (Smith Indic M.B.) XC ff. 1–2. On f. 2v is written: śloka 150.
Ff. 1r–1v. Mean motion tables of the Lord of the Year, Epact, Moon, Moon's Anomaly, Mars, Mercury, Jupiter, Venus, Saturn, Lunar Node, and Precession for 0 to 9 years and for 1 to 9 groups of 10 years; computed (where applicable) according to the *Mahādevī*. Yearly motions for the Moon, Lunar Node, and Precession are, respectively: 2,12;46,40,32°; – 19;21,34,0,6°; and 0;1, 0,0°. The epoch positions, given as the entry for 0 years, indicate as epoch date Sunday, 31 March 1678 in the Julian calendar.

4946 (Smith Indic M.B.) XCI ff. 1–2.
Ff. 1r–1v. Table 7 of the *Candrārkī* for 1 to 58 years.
Ff. 2r–2v. Table 6 of the *Candrārkī* for 0° to 29° horizontal, and for 7 to 12 zodiacal signs vertical. See 4949.

4946 (Smith Indic M.B.) XCII ff. 1–2 and 5. On f. 1r is written: śloka 175.
Ff. 1r–2v. Table 6 of the *Candrārkī* for 0° to 29° horizontal, and for 1 to 12 zodiacal signs vertical.
Ff. 2v and 5r. Part of table 1 of the Anonymous of 1704: for 31 to 61 days and for 304 to 365 days.
F. 5v. Text on aspect of Saturn.

4946 (Smith Indic M.B.) XCIII ff. 1–2.
F. 1r. Table 4 of the Anonymous of 1704 for 1 to 60 ghaṭikās.
F. 1v. Table 5 of the Anonymous of 1704 for 0 to 42 years.
F. 2r. Table 2 of the Anonymous of 1704 for 0 to 88 years.
F. 2v. Blank.

4946 (Smith Indic M.B.) XCIV 1 f.
F. 1r. Table 8 of the *Mahādevī*.
F. 1v. Blank.

4946 (Smith Indic M.B.) XCV f. 1.
Ff. 1r–1v. Text on daśās.

4946 (Smith Indic M.B.) XCVI f. 1. On f. 1r, in the right-hand margin, is written in European numerals: 6589.
Ff. 1r–1v. Tables of the longitudes of the five planets and the Lunar Node for Saṃ. 1852, Śaka 1717 (A.D. 1795); computed according to the Anonymous of 1704.

4946 (Smith Indic M.B.) XCVII 1 f. On f. 1v is

written: śloka 75.
F. 1r. Table 3 of the *Mahādevī*.
F. 1v. Blank.

4946 (Smith Indic M.B.) XCVIII 1 f.
F. 1r. Multiplication tables for 1;48 by 1 to 60, etc.
Ff. 1r–1v. Unidentified astrological text.

4946 (Smith Indic M.B.) XCIX 1 f.
Ff. 1r–1v. Unidentified astrological text.

4946 (Smith Indic M.B.) C ff. 1–2.
F. 1r. Table 14 of the Anonymous of 1741.
Ff. 1v–2v. Unidentified astrological text.

4946 (Smith Indic M.B.) CI f. 1. On f. 1r is written: śloka 800.
Ff. 1r–1v. Tables of the longitudes of the five planets and the Lunar Node for Śaka 1769 (A.D. 1847); computed according to the Anonymous of 1704.

4946 (Smith Indic M.B.) CII 3 ff., each numbered 1.
Ff. 1r–1v. Table of the mean motion of the Moon and the Lunar Anomaly for 49 to 135 years. The respective yearly parameters are 2,12;46,40,32° and 1,32;6,9,13°.
Ff. 2r–2v. Table of the Epact and tithikendra for Śaka 1749 to 1825 (A.D. 1827 to 1903). The respective yearly parameters are 11;3,53,22,40 tithis and 7;40,30,44,20 tithis.
Ff. 3r–3v. Table of the Lord of the Year for Śaka 1749 to 1860 (A.D. 1827 to 1938). The yearly parameter is 1;15,31,17,17 days.

4946 (Smith Indic M.B.) CIII f. 1.
Ff. 1r–1v. Table of oblique ascensions for 0° to 29° horizontal, and for 1 to 12 zodiacal signs vertical. The vernal equinox is at Pisces 8°.

4946 (Smith Indic M.B.) CIV 1 f.
F. 1r. Table of the beginnings of the nakṣatrapādas of the 27 nakṣatras.
F. 1v. Short astrological text.

4946 (Smith Indic M.B.) CV 1 f.
Ff. 1r–1v. Table related to oblique ascensions for finding the ascendent. Precession is stated to have been 22°.

4949 (Smith Indic 19) 2 ff. On f. 1r, in the upper margin, is written: śloka 50. The rest of this manuscript is 4946 XCI.
Ff. 1r–1v. Table 6 of the *Candrārkī* from Aries 0° to Libra 29°.
F. 2r. Table 7 of the *Candrārkī* for years 59 to 84.
F. 2v. Blank.

4950 (Smith Indic 17) 12 ff. On f. 1r, in the upper margin, is written: śloka 3000.
F. 1r. Blank.
F. 1v. Introductory verses of the *Khecaradīpikā*.
F. 1v. Table 1 of the *Khecaradīpikā*.
F. 2r. Explanatory text.
F. 2v. Table 2 of the *Khecaradīpikā*.
Ff. 3r–4v. Table 3 of the *Khecaradīpikā* for N = 0–11.
Ff. 5r–7v. Table 4 of the *Khecaradīpikā* for N = 0–11.
Ff. 8r–9v. Table 5 of the *Khecaradīpikā* for N = 0–11.
Ff. 10r–11v. Table 6 of the *Khecaradīpikā* for N = 0–11.
Ff. 12r–12v. Table 7 of the *Khecaradīpikā* for N = 0–5.

4951 (Smith Indic 138 f. 27v). See 4855.

4952 (Smith Indic 29) 6 ff. On f. 6r is written: śloka 300. On f. 6v, at the bottom, is written: iti likhitaṃ malūkacandreṇa.
Ff. 1r–2v. Table 2 of the *Candrārkī* for 0 to 365 days.
Ff. 2v–3r. Table 3 of the *Candrārkī* for 0 to 60 ghaṭikās.
Ff. 3r–6r. Table 4 of the *Candrārkī* for 0° to 359°.
Ff. 6r–6v. Several verses beginning:

śrīgaṇeśāya namaḥ
natvā vallabhanandanaṃ tad⟨a⟩nugopālāṅghripadmadvayaṃ
jñātvā śrīguruvākyato hy aharniśaṃ⟨—⟩dyum evādhunā /
siddhānteṣu yathoktakhecaravidhiḥ ⟨su⟩spaṣṭakoṣṭho muhur
madhyaspaṣṭavibhāgato grahagaṇāt kurv⟨e⟩ dinaughād aham // 1 //

and ending: iti brahmatulyasāraṇīślokāḥ //

candrarāśau kalāḥ sarve dvādaśair bhāgam āharet /
yatrodvāhe śubhe kārye candrāvasthā parityajet //
1 2 3 4 5 6 7 8
pravāsanaṣṭāmṛtā jayākhyā hāsyāratikrīḍitasuptab-
9
huktāḥ /
10 11 12
jarāṅkayākampitasusthit⟨āś⟩ ca meṣādimukhyā himagor avasthāḥ //.

4955 (U. Penn. 1913 b and c). See 4798.

5107 (Harvard 61) 23 ff.
F. 1r. Blank.
F. 1v. Table 1 of the *Jayavinodasāraṇī* for Śaka 1657, 1687, 1717, 1747, 1777, and 1807 (A.D. 1735, 1765, 1795, 1825, 1855, and 1885).
Ff. 2r–2v. Table 2 of the *Jayavinodasāraṇī* for 1 to 30 years.
Ff. 3r–6r. Table 3 of the *Jayavinodasāraṇī* for 0 to 388 days.
F. 6v. Blank.
Ff. 7r–8v. Table 4 of the *Jayavinodasāraṇī* for 0 to 29 days horizontal, 0 to 59 ghaṭikās vertical.
Ff. 9r–23v. Table 5 of the *Jayavinodasāraṇī* for 0 to 388 days horizontal, 0 to 29 ghaṭikās vertical. The title of this table is: yoge ravikendraphalāni jayavinode.

5108 (U. Penn. 1824) ff. 2–10.
Ff. 2r–2v. Table 1 of the Anonymous of 1520 for 1 to 67 days.
F. 2v. Table 2 of the Anonymous of 1520 for periods 17 to 33; dhruva and kṣepaka given.

Ff. 3r–3v. Table 3 of the Anonymous of 1520 for 1 to 67 days (and so by moving the sexagesimal place, for 1,0 to 1,10,0 days).
F. 3v. Table 4 of the Anonymous of 1520 for periods 17 to 33; dhruva and kṣepaka given.
Ff. 4r–4v. Table 5 of the Anonymous of 1520 for 1 to 67 days.
F. 4v. Table 6 of the Anonymous of 1520 for periods 17 to 33; dhruva and kṣepaka given.
Ff. 5r–5v. Table 9 of the Anonymous of 1520 for 1 to 67 days.
F. 5v. Table 10 of the Anonymous of 1520 for periods 17 to 33; dhruva and kṣepaka given.
Ff. 6r–6v. Table 11 of the Anonymous of 1520 for 1 to 67 days.
F. 6v. Table 12 of the Anonymous of 1520 for periods 17 to 33; dhruva and kṣepaka given.
Ff. 7r–7v. Table 13 of the Anonymous of 1520 for 1 to 67 days.
F. 7v. Table 14 of the Anonymous of 1520 for periods 17 to 33; dhruva and kṣepaka given.
Ff. 8r–8v. Table 15 of the Anonymous of 1520 for 1 to 67 days.
F. 8v. Table 16 of the Anonymous of 1520 for periods 17 to 33; dhruva and kṣepaka given.
Ff. 9r–9v.Table 17 of the Anonymous of 1520 for 1 to 67 days.
F. 9v. Table 18 of the Anonymous of 1520 for periods 17 to 33; dhruva and kṣepaka given.
Ff. 10r–10v. Table 7 of the Anonymous of 1520 for 1 to 67 days.
F. 10v. Table 8 of the Anonymous of 1520 for periods 17 to 33; dhruva and kṣepaka given.

In the left-hand margin of f. 10r is written the mean daily motion of the Lunar Node: – 0;3,10,48,25,14,47, 28°.

5125 (Smith Indic 138 f. 27r). See 4855.

5171 (U. Penn. 702) 16 ff. On f. 1r is written: sāriṇī / naṃ 69 pa⟨ttra⟩ 16.
Ff. 1v–7r. Table 1 of the *Tithicintāmaṇi* for 0 to 400 tithis.
Ff. 7r–10r. Table 2 of the *Tithicintāmaṇi* for 0 to 390 nakṣatras.
Ff. 10v–16r. Table 3 of the *Tithicintāmaṇi* for 0 to 420 yogas.
F. 16r. Table 5 of the *Tithicintāmaṇi* for 1 to 12 zodiacal signs.
F. 16r. Table 4 of the *Tithicintāmaṇi* for 1 to 27 nakṣatras.

5172 (U. Penn. 1877) 2 ff.
Ff. 1r–1v. Table 19 of the Anonymous of 1520 for 0° to 90°.
Ff. 2r–2v. Table 20 of the Anonymous of 1520 for 0° to 90°.

5177 (Smith Indic 28) ff. 8–23. On f. 23v, at the top, is written: ślo 400, and, at the bottom: saṃvat 1846 nā jyeṣṭha su da 9 ravīvāre tithīsāraṇī graṃtha lakhyate paṃḍā lālajī premajī leṣyake gurujī śrī amarasījī nā. The date corresponds to Sunday, 31 May 1789.
F. 8r. Table 1 of the Anonymous of 1741.
F. 8v. Table 2 of the Anonymous of 1741.
F. 9r. Table 3 of the Anonymous of 1741.
F. 9v. Table 4 of the Anonymous of 1741.
F. 10r. Table 5 of the Anonymous of 1741.
F. 10v. Table 6 of the Anonymous of 1741.
F. 11r. Table 7 of the Anonymous of 1741.
F. 11v. Table 8 of the Anonymous of 1741.
F. 12r. Table 9 of the Anonymous of 1741.
F. 12v. Table 10 of the Anonymous of 1741.
F. 13r. Table 11 of the Anonymous of 1741 for 0 to 36 periods of 10 days.
F. 13v. Table 12 of the Anonymous of 1741 for 0 to 41 periods of 9 days.
F. 14r. Table 13 of the Anonymous of 1741 for 0 to 43 periods of 9 days.
F. 14v. Table 14 of the Anonymous of 1741 for 1 to 27 nakṣatras.
F. 14v. Table 15 of the Anonymous of 1741 for 1 to 12 zodiacal signs.
Ff. 15r–17v. Table 16 of the Anonymous of 1741 for 0 to 27 horizontal, and for 0 to 59 vertical.
Ff. 18r–20v. Table 17 of the Anonymous of 1741 for 0 to 27 horizontal, and for 0 to 59 vertical.
Ff. 21r–23v. Table 18 of the Anonymous of 1741 for 0 to 29 horizontal, and for 0 to 59 vertical.

5178 (Smith Indic 35) 11 ff.
F. 1r. Blank.
Ff. 1v–4v. Table 1 of the *Candrārkī* for 0 to 365 days.
Ff. 4v–7v. Table 2 of the *Candrārkī* for 0 to 365 days.
F. 8r. Table 3 of the *Candrārkī* for 1 to 60 ghaṭikās.
Ff. 8v–11v. Table 4 of the *Candrārkī* for 0° to 359°.

5179 (Smith Indic 46) ff. 2–10. On f. 2r, at the top, is written: śloka 250.
Ff. 2r–4r. Table 1 of the *Candrārkī* for 104 to 365 days.
Ff. 4v–7v. Table 2 of the *Candrārkī* for 0 to 365 days.
F. 8r. Table 3 of the *Candrārkī* for 1 to 60 ghaṭikās.
Ff. 8v–10r. Table 4 of the *Candrārkī* for 0° to 239°.
F. 10v. Skeleton of continuation of table 4 of the *Candrārkī;* not filled in.

5180 (Smith Indic 82) 3 ff.
Ff. 1r–1v. Table 11 of the Anonymous of 1741 for 0 to 37 periods of 10 days.
Ff. 2r–2v. Table 12 of the Anonymous of 1741 for 0 to 41 periods of 9 days.
Ff. 3r–3v. Table 13 of the Anonymous of 1741 for 0 to 43 periods of 9 days.

5181 (Smith Indic 87) 1 f. On f. 1v, in the left-hand margin, is written in European numerals: 6578.

Ff. 1r–1v. A Nakṣatracaraṇapraveśapattra for each of the 4 caraṇas in the 27 nakṣatras.

5183 (U. Penn. 1902) 3 ff.
Ff. 1r–3r. Beginning of Divākara's *Makarandavivaraṇa.*
F. 3v. Blank.

5187 (Smith Indic 129 B ff. 22–30). See 4801.

5188 (Smith Indic 129 B ff. 2–21v). See 4801.

BRAHMATULYASĀRAṆĪ

I. These tables utilize the parameters of the *Brahmatulya* (otherwise known as the *Karaṇakutūhala*) of Bhāskara II, who was born in 1114. Their epoch, as indicated below, is 23 February 1183, though they themselves were probably written later.

II. Manuscripts:

*Poleman 4735 (Smith Indic 45) ff. 2–10 and 12–17. Saṃ. 1855, Śaka 1720 = A.D. 1798.
*Poleman 4876 (Smith Indic 43) 13 ff.
*Poleman 4946 (Smith Indic MB) LVIII ff. 1–5.
*Poleman 4946 (Smith Indic MB) LXII f. 11.
*Poleman 4952 (Smith Indic 29) f. 6.

III. Tables:

The mean motion tables are for 1 to 30 days; 1 to 12 "months" (1 "month" = 30 days); 1 to 20 "years" (1 "year" = 360 days); and 1 to 60 periods of 20 "years" (the epoch position is added to the first of these entries), The mean daily motions are:

Saturn	0;2,0,23,3,30°
Jupiter	0;4,59,8,54,0
Mars	0;31,26,28,9,50
Sun	0;59,8,10,12,40
Venus' conjunction	1;36,7,43,49,50
Mercury's conjunction	4;5,32,21,1,0
Moon	13;10,34,52,31,50
Lunar Apogee	0;6,40,53,50,10
Lunar Node	– 0;3,10,48,25,30

The epoch positions indicate a date of 23 February 1183:

Saturn	123;43,17	140	–16
Jupiter	64;0,51	67	–3
Mars	231;24,21	270	–39
Sun	329;13,0	341	–12
Venus' conjunction	258;5,55	(Venus) 311	
Mercury's conjunction	81;14,30	(Mercury) 359	
Moon	329;5,50	c.335	–5
Lunar Apogee	135;12,59		
Lunar Node	82;34,51	c.78	–5

1. Mean motion table for the Sun.
Manuscripts: 4735 f. 2; 4946 LVIII f. 1.

2. Mean motion table for the Moon.
Manuscripts: 4735 f. 2v; 4946 LVIII f. 1v.

3. Mean motion table for the Lunar Apogee.
Manuscripts: 4735 f. 3; 4946 LVIII f. 2.

4. Mean motion table for the Lunar Node.
Manuscripts: 4735 f. 3v; 4946 LVIII f. 2v.

5. Mean motion table for Mars.
Manuscripts: 4735 f. 4; 4946 LVIII f. 3.

6. Mean motion table for Mercury's conjunction.
Manuscripts: 4735 f. 4v; 4946 LVIII f. 3v.

7. Mean motion table for Jupiter.
Manuscripts: 4735 f. 5; 4946 LVIII f. 4.

8. Mean motion table for Venus' conjunction.
Manuscripts: 4735 f. 5v; 4946 LVIII f. 4v.

9. Mean motion table for Saturn.
Manuscripts: 4735 f. 6; 4946 LVIII f. 5.

10. Table of the equation of the center of the Sun for 0° to 90°; also given are the differences between successive equations, and the increment to (or decrease from) the mean daily motion. All the tables of the equation of the center in the *Brahmatulyasāraṇī* are set up in this way. The maximum equation is 2;10,54°; the maximum increment 0;2,20°.
Manuscripts: 4735 f. 6v; 4876 ff. 1v–2.

11. Table of the equation of the center of the Moon for 0° to 90°. The maximum equation is 5;2,31°; the maximum increment 1;8,15°.
Manuscripts: 4735 f. 7; 4876 ff. 2–2v.

12. Table of Solar declination and of lunar latitude for 0° to 90°. The obliquity of the ecliptic equals 24;0°; the maximum lunar latitude 5;42,51°.
Manuscripts: 4735 f. 7v; 4876 f. 1; 4946 LVIII f. 5v.

13. Table of precession (bhaumamandocca) for 1 to 45 periods of 10 years. The rate of precession is 0;0,54° per year.
Manuscripts: 4735 f. 7v; 4876 ff 1–1v.

14. Table of the equation of the center of Mars for 0° to 90°. The maximum equation is 11;12,53°; the maximum increment 0;6,0°.
Manuscripts: 4735 f. 8; 4876 ff 2v–3.

15. Table of the equation of the conjunction of Mars for 0° to 180°; also included are the differences between successive equations, and a third function which must, in some unexplained manner, be related to the planet's daily progress. All the tables of the equations of the conjunction of the planets are set up in this way. The third function reaches a maximum at an anomaly of 1°, a minimum at one of 180°.

	maximum	minimum
Saturn	132;58,49	107;0,0
Jupiter	142;58,4	97;0,0
Mars	200;52,10	39;0,0
Venus	206;54,57	33;0,0
Mercury	163;56,46	76;0,0

The maximum equation of the conjunction of Mars is 41;17,59° at 130°.

Manuscripts: 4735 ff 8v–9v; 4876 ff 6–7v.

16. Table of the equation of the center of Mercury. The maximum equation is 6;3,38°; the maximum increment 0;6,0°.

Manuscripts: 4735 f. 10; 4876 ff 3–3v.

17. Table of the equation of the conjunction of Mercury. The maximum equation is 21;36,48° at 110°.

Manuscripts: 4735 f. 10v; 4876 ff. 8–9; 4946 LXII ff. 11–11v (= 4735 ff 11–11v.)

18. Table of the equation of the center of Jupiter. The maximum equation is 5;15,47°; the maximum increment 0;0,25°.

Manuscripts: 4735 f. 12; 4876 ff 4–4v.

19. Table of the equation of the conjunction of Jupiter. The maximum equation is 10;59,1° at 100°.

Manuscripts: 4735 ff 12v–13v; 4876 ff. 9v–10v.

20. Table of the equation of the center of Venus. The maximum equation is 1;31,50°; the maximum increment 0;1,45°.

Manuscripts: 4735 f. 14; 4876 ff 4v–5.

21. Table of the equation of the conjunction of Venus. The maximum equation is 46;30,28° at 140°.

Manuscripts: 4735 ff 14v–15v; 4876 ff 11–12.

22. Table of the equation of the center of Saturn. The maximum equation is 7;38,35°; the maximum increment 0;0,10°.

Manuscripts: 4735 f. 16; 4876 ff 5–5v.

23. Table of the equation of the conjunction of Saturn. The maximum equation is 6;10,7° at 100°.

Manuscripts: 4735 ff 16v–17v; 4876 ff 12v–13v.

THE *MAHĀDEVĪ* OF MAHĀDEVA

I. Life of Mahādeva:

Mahādeva's genealogy is given in verses 41–43 of the *Grahasiddhi*, unfortunately rather corruptly preserved in the copies available to me; I restore them tentatively as follows:

> īśvarakauverayajñadaśamadhye sāmagāgrajanmāsīt /
> śrībhogadevanāmā gautamagotraḥ sa daivajñaḥ // 41 //
> tatputro mādhavadaivavit tatputraḥ padmanābhadaivajñaḥ/
> tajjo godāvarīśīrṇe (?) paraśurāmadaivajñaḥ// 42 //
> tajjamahādevena grahasiddheḥ siddhaye yatitam/
> tacchreyasā bhavānībhavaś ca bhavatu dvijaprajāvṛddhyai// 43 //

Mahādeva was the son of the astrologer Paraśurāma, who lived on the bank of the Godāvarī (godāvarītīre?); Paraśurāma's father was the astrologer Padmanābha, the son of the astrologer Mādhava, the son of the astrologer Bhogadeva of the Gautamagotra, a follower of the *Sāmaveda* and performer of sacrifices to Īśvara and Kauvera. Mahādeva's date is indicated by the epoch of the *Mahādevī*: Sunday, 28 March 1316.

In the first verse of the *Grahasiddhi* Mahādeva pays his respects to the gods and to his teachers. The second verse names some of his famous predecessors among Indian astronomers.

> siddhiṃ karotīśvarakendrabhorvīnadīḥ śivau kṣetrapavāggurūṃś ca /
> cakreśvarārabdhanabhaścarāṇāṃ siddher mahādeva ṛṣīṃś ca natvā / 1 /
> paitāmahāryabhaṭajiṣṇujabhāskarādisiddhāntabhedakaraṇair nitarām agādhe /
> saṃkhyārṇave khacarakarmajale nimagnajyotirvidāṃ prataraṇāya dṛḍhorunāvaḥ / 2 /

Despite the fact that Mahādeva apparently says his father lived on the banks of the Godāvarī in the Deccan (perhaps near Aurangabad?), he himself seems to have come from Rajasthan or Gujarat, where most of his manuscripts are found. His identifiable commentators are Nṛsiṃha (1528) of Nandipura, (modern Nandod on the Narmada) Dhanarāja (1635) of Padmāvatī in Mārwār, and Mādhava.

II. Manuscripts of the *Mahādevī*:

Anup 4960. 34 ff. Śaka 1467 = A.D. 1545.
Goṇḍal 257. 75 ff. Saṃ. 1675 = A.D. 1618.
LDI 8825. 1 f. Saṃ. 1695 = A.D. 1638 (*Grahasiddhi*)
Baroda 689. 32 ff. Saṃ. 1719 = A.D. 1662 (with ṭīkā of Dhanarāja)
BORI 340 of 1879/80. 38 ff. Saṃ. 1722 = A.D. 1665 (with ṭīkā of Dhanarāja)
*Poleman 4763a (Smith Indic 41) 128 ff. Śaka 1588 = A.D. 1666.
LDI 7129. 48 ff. Sam. 1729 = A.D. 1672 (with ṭīkā of Dhanarāja)
BORI 497 of 1892/95. 25 ff. Saṃ. 1734 = A.D. 1677. (with ṭīkā of Dhanarāja)
Goṇḍal 263. 150 ff. Saṃ. 1744 = A.D. 1687.
LDI 5132. 35 ff. Saṃ. 1754 = A.D. 1697 (with ṭīkā of Dhanarāja).
BORI 845 of 1887/91. 38 ff. Saṃ. 1761 = A.D. 1704 (with ṭīkā of Dhanarāja).
Goṇḍal 261. 159 ff. Saṃ. 1773, Śaka 1638 = A.D. 1716.
Tod 24. 63 ff. Saṃ. 1776 = A.D. 1719.
LDI 8877. 41 ff. Saṃ. 1779 = A.D. 1722 (with ṭīkā of Dhanarāja)
VVRI 1459. 5 ff. Saṃ. 1786 = A.D. 1729.
Tod 45. 77 ff. Saṃ. 1789, Śaka 1654 = A.D. 1732.
Goṇḍal 260. 91 ff. Saṃ. 1803 = A.D. 1746.
*Poleman 4944 (Smith Indic 172). 151 ff. Saṃ. 1814 = A.D. 1757
RORI 4741. 126 ff. Saṃ. 1814 = A.D. 1757.
LDI 7028. 3 ff. Saṃ. 1822 = A.D. 1765.
RORI 670. 1 f. Saṃ. 1826 = A.D. 1769.
Goṇḍal 264. 142 ff. Saṃ. 1841 = A.D. 1784.
Goṇḍal 265. 18 ff. Saṃ. 1845 = A.D. 1788.

LDI 7513. 76 ff. Saṃ. 1846 = A.D. 1789.
BORI 429 of 1895/98. 75 ff. Saṃ. 1847 = A.D. 1790.
RORI 3759. 152 ff. Saṃ. 1848 = A.D. 1791.
RORI 4786. 152 ff. Saṃ. 1849 = A.D. 1792.
LDI 7412. 29 ff. Saṃ. 1852 = A.D. 1795 (with ṭīkā of Dhanarāja).
Goṇḍal 262. 50 ff. Saṃ. 1856, Śaka 1721 = A.D. 1799.
Goṇḍal 259. 72 ff. Saṃ. 1857 = A.D. 1800.
RORI 3749. 92 ff. Saṃ. 1861 = A.D. 1804.
Goṇḍal 256. 151 ff. Saṃ. 1890 = A.D. 1833.
Goṇḍal 255. 26 ff. Saṃ. 1902 = A.D. 1845 (with ṭīkā of Jainasādhu = Dhanarāja)
Goṇḍal 71. 2 ff. Saṃ. 1921 = A.D. 1864. (*Grahasiddhi*).
Goṇḍal 258. ff. 112–142. Saṃ. 1933 = A.D. 1876.
Goṇḍal 128c and 128e. ff 2–76 and 87–111. Saṃ. 1935 = A.D. 1878.
Adyar Index 1957.
Anup 4957. 192 ff.
Anup 4958. 61 ff.
Anup 4959. 158 ff.
Anup 4961. 32 ff.
ASBombay 254. 30 ff. (with ṭīkā of Dhanarāja).
Baroda 3111. 2 ff. (*Grahasiddhi*)
Baroda 9546. 5 ff.
CP, Hiralal 6445. (from Gaṛhākoṭā in Saugor District)
*Florence 458. 2 ff. (*Grahasiddhi*)
Jaipur. 76 ff.
Jaipur. 4 ff.
Kotah 167. 154 pp.
LDI 6646. 150 ff.
LDI 7047. 3 ff. (*Grahasiddhi*)
LDI 7108. 78 ff.
LDI 7127. 3 ff. (*Grahasiddhi*)
LDI 7366. 8 ff.
LDI 7677. 3 ff. (*Grahasiddhi*)
Paris BN 1005 (Sans. Dév. 331–340) XV.
PL, Buhler 323. 6 ff.
PL, Buhler 431. (*Laghumahādevī*).
*Poleman 4763 (Smith Indic 192). 29 ff.
*Poleman 4764 (Smith Indic 80). 9 ff.
*Poleman 4765 (Smith Indic 22). 3 ff.
*Poleman 4767 (Smith Indic 18). ff. 1 and 3–7.
*Poleman 4769 (Smith Indic 147, pt. 1). 30 ff.
*Poleman 4770 (Smith Indic 179). f. 3. (*Grahasiddhi*)
*Poleman 4771 (Smith Indic 91). 8 ff.
*Poleman 4772 (Smith Indic 193). ff. 49–75.
*Poleman 4773 (Smith Indic 147, pt. 2). ff. 31–60 and 1 f.
*Poleman 4864 (Smith Indic 84). 2 ff.
*Poleman 4866 (Smith Indic 101). 8 ff.
*Poleman 4867 (Smith Indic 105). 46 ff.
*Poleman 4868 (Smith Indic 140). 177 ff.
*Poleman 4870 (Smith Indic 191). 2 ff.
*Poleman 4946 (Smith Indic MB) XXII. 1 f.
*Poleman 4946 (Smith Indic MB) XXV. 1 f.
*Poleman 4946 (Smith Indic MB) XXIX. 1 f.
*Poleman 4946 (Smith Indic MB) XXX. 1 f.
*Poleman 4946 (Smith Indic MB) LIX. 1 f.
*Poleman 4946 (Smith Indic MB) LXI. f. 91
*Poleman 4946 (Smith Indic MB) LXIII. 4 ff.
*Poleman 4946 (Smith Indic MB) LXIV. 1 f.
*Poleman 4946 (Smith Indic MB) LXV. f. 26.
*Poleman 4946 (Smith Indic MB) LXXVII. f. 1.
*Poleman 4946 (Smith Indic MB) XCIV. f. 1.
*Poleman 4946 (Smith Indic MB) XCVII. 1 f.
RORI 625. 3 ff.
RORI 2004. 39 ff.
RORI 3743. 76 ff.
RORI 3778. 5 ff. (*Grahasiddhi* and *Labdhaśeṣa*).
RORI 3680. 91 ff.
RORI 4892. 3 ff. (*Grahasiddhi*).

III. Tables:

The structure and significance of these tables have been discussed by O. Neugebauer and the present writer in a previous paper (*Proc. Amer. Philos. Soc.* **111**, 1967, 69–92); here it is necessary to give only the references to the manuscripts which contain each table and to the parameters which are reiterated by Mahādeva's imitators and successors.

A. Mean motion tables.

1. Epact. Parameter: 11;3,53,22,40 tithis per year.
Manuscripts: 4764 ff. 2–2v; 4767 ff. 1–1v; 4771 ff. 1–1v; 4866 ff. 1–1v; 4867(c) ff. 1–1v; 4868B ff. 4–4v; 4946 XXIX f. 1v; 4946XXX f. 1; 4946LXIII f. 1; 4946 LXXVII f. 1v.

2. Lord of the year. Parameter: 1;15,31,17,17 days per year.
Manuscripts: 4764 ff. 1–1v; 4771 ff. 2–2v; 4866 ff. 2–2v; 4867(c) ff. 2–2v; 4868B ff. 5–5v; 4946XXX f. 1v; 4946 LXVII f. 1.

2a. Moon. Parameter: 2,12;46,40,32,0° per year.
Manuscripts: 4764 ff. 3–3v.

3. Mars. Parameter: 3,11;24,9,20,6° per year.
Manuscripts: 4764 ff. 4–4v; 4765 f. 1; 4767 ff. 3–3v; 4771 ff. 3–3v; 4866 ff. 3–3v; 4867(c) ff. 3–3v; 4868B ff. 6–6v; 4946LXIV f. 1; 4946XCVII f. 1.

4. Mercury. Parameter: 54;45,19,2,12° per year.
Manuscripts: 4764 ff. 5–5v; 4765 f. 1v; 4767 ff. 4–4v; 4771 ff. 4–4v; 4866 ff. 4–4v; 4867(c) ff. 4–4v; 4868B ff. 7–7v.

5. Jupiter. Parameter: 30;21,6,30,42° per year.
Manuscripts: 4764 ff. 6–6v; 4765 f. 2; 4767 ff. 5–5v; 4771 ff. 5–5v; 4866 ff. 5–5v; 4867(c) ff. 5–5v; 4868B ff. 8–8v.

6. Venus. Parameter: 3,45;11,53,48,36° per year.
Manuscripts: 4764 ff. 7–7v; 4765 f. 2v; 4767 ff. 6–6v; 4771 ff. 6–6v; 4866 ff. 6–6v; 4867(c) ff. 6–6v; 4868B ff. 9–9v.

7. Saturn. Parameter: 12;12,51,25,12° per year.
Manuscripts: 4764 ff. 8–8v; 4765 f. 3; 4767 ff. 7–7v; 4771 ff. 7–7v; 4866 ff. 7–7v; 4867(c) ff. 7–7v; 4868B ff. 10–10v.

8. Lunar Node. Parameter: 5,40;38,26,0,6° (= −19; 21,33,59,54°) per year.

Manuscripts: 4764 ff. 9–9v; 4765 f. 3v; 4771 ff. 8–8v; 4866 ff.8–8v; 4867(c) ff. 9–9v; 4868B ff. 11–11v; 4946 XCIV f. 1.

8a. Bījas.

Manuscripts: 4770 f. 3v.

B. True Longitudes.

9. Mars.

Manuscripts: 4763 ff. 1–15v (N = 0 to 59); 4763a ff. 9–18, 28, and 30 (N = 16 to 35, 54 to 55, and 58 to 59); 4769 ff. 1–30 (N = 0 to 59); 4867(a) 30 ff. (N = 0 to 59); 4868A ff. 1–18 (N = 0 to 35); 4868C ff. 1–30 (N = 0 to 59); 4944 ff. 1–30 (N = 0 to 59).

10. Mercury.

Manuscripts: 4763 ff. 16–29v (N = 0 to 55); 4763a ff. 4 and 6–30 (N = 6 to 7 and 10 to 59); 4773 (= 4769 pt. 2) ff. 31–60 (N = 0 to 59); 4773 unnumbered f. (N = 30); 4867(b) ff. 2–9 (N = 2 to 17); 4868C ff. 31–60 (N = 0 to 59); 4944 ff. 31–60 (N = 0 to 59); 4946 XXII 1 f. (N = 56 to 57). 4946 XXV 1 f. (N = 1); 4946 LXIII ff. 2–4 (N = 3, 13, and 15); 4946 LXV f. 26 (N = 18 to 19).

11. Jupiter.

Manuscripts: 4763a ff. 1–30 (N = 0 to 59); 4868C ff. 61–90 (N = 0 to 59); 4944 ff. 61–90 (N = 0 to 59); 4946 LIX 1 f. (N = 10).

12. Venus.

Manuscripts: 4763a ff. 1–30 (N = 0 to 59); 4772 ff. 49–60 (N = 12 to 59); 4868C ff. 91–120 (N = 0 to 59); 4944 ff. 91–120 (N = 0 to 59).

13. Saturn.

Manuscripts: 4763a ff. 1–30 (N = 0 to 59); 4772 ff. 61–75 (N = 0 to 59); 4868C ff. 121–150 (N = 0 to 59); 4944 ff. 121–150 (N = 0 to 59); 4946 LXI f. 91 (N = 58 to 59).

C. True Longitudes for Specific Years.

1. Saṃ. 1842 = A.D. 1785: 4870. 2 ff.

2. Saṃ. 1857, Śaka 1722 = A.D. 1800: 4864. f. 1.

3. Saṃ. 1859, Śaka 1724 = A.D. 1802: 4864 f. 2.

D. Associated Texts.

1. Commentary: 4763a Mercury ff. 31–31v; 4768 5 ff.; 4770 f. 3.

2. Mean motion tables for 90-year periods: 4946 XC ff. 1–2.

3. Mean motion tables for 900-year periods: 4946 LXVIII f. 1.

ANONYMOUS OF 1461

I. Manuscripts.

*Poleman 4859 + 4951 (Smith Indic 138 ff. 24–26 + 27v)

II. Tables.

1. Table of lengths of daylight in ghaṭikās for 1 to 366 days. Longest day is 33;12 ghaṭikās on days 73 and 74; shortest day is 26;48 ghaṭikās on day 257. This implies a latitude of about 23°—i.e., approximately that of Ujjain. It also indicates about 15° of precession; at the common rate of 1° every 66 years, this places the text about 1000 years after the fixing of the Indian zodiac —i.e., c. 1500 A.D.

Manuscripts: 4859 (Smith Indic 138 ff. 24-26v)

2. Table of planetary positions entitled spaṣṭadhruvaka. These longitudes indicate as date 13 May 1461.

	Sidereal		Tropical	Sidereal-Tropical
Sun	40;2,2°		61°	−21°
Moon	82;56,2		106	−23
Lunar apogee	240;30,0			
Lunar node	2;20,2		c105!	−103!
Marș	141;44,0		171	−29
Mercury (conj)	195;54,0	(Mercury)	83	
Jupiter	218;36,0		235	−16
Venus (conj)	248;44,0	(Venus)	35	
Saturn	285;24,0		302	−17

Manuscripts: 4859 (Smith Indic 138 f. 26v)

3. Table of cālanas, whose values are the planets' mean motions for 1/27 of a year (= 13;31,41,10 days); this is called a cālana by the text in 4951.

Sun	13;20,0°	Saturn	0;27,8
Mars	7;5,21	Lunar Node	5,59;16,59
Mercury	42;1,40	Lunar Node	−0;43,1
Jupiter	1;7,27	Moon	2,58;15,3
Venus	8;20,27	Lunar Apogee	1;30,22

Manuscripts: 4859 (Smith Indic 138 f. 26v); 4951 (Smith Indic 138 f. 27v).

4. Table of cālanas of the nakṣatras.

Aśvinī	0,40	Maghā	2,16
Bharaṇī	2,9	Pūrvaphālgunī	2,44
Kṛttikā	2,13	Uttaraphālgunī	2,43
Rohiṇī	1,4	Hasta	2,43
Mṛgaśiras	0,53	Citra	2,43
Ārdrā	1,6	Svāti	2,24
Punarvasu	0,53	Viśākhā	2,13
Puṣya	1,23	Anurādhā	1,43
Āśleṣā	2,13	Jyeṣṭhā	0,36

Manuscripts: 4859 (Smith Indic 138 f. 26v).

THE *MAKARANDA* OF MAKARANDA

I. Makaranda's Life:

On Makaranda himself our only source of information is the first verse of his work:

śrīsūryasiddhāntamatena samyag viśvopakārāya gurūpadeśāt /
tithyādipattraṃ vitanoti kāśyām ānandakando makarandanāmā //

This tells us that he wrote the *Makaranda* at Kāśī (Benares) on the basis of the *Sūryasiddhānta*. Makaranda's date is indicated by the epoch of his tables: 1478.

The *Makaranda* was extremely popular; there are commentaries and derivative texts by: Harikarṇa (1610) of Hisāranagara (modern Hissar in the Panjab), Viśvanātha (fl. 1612–1630) of Benares, Divākara (1627), Puruṣottama (1631), Kṛpārāma Miśra (1809) of Ahmadabad, Jīvanātha (fl. 1823) of Patna, Nīlāmbara Jhā (nineteenth century) of Koilakh in Mithilā, Kṣemaṅkara, Gokulanātha, Cūḍāmaṇi Cakravartin, Ḍhuṇḍhirāja, Nīlakaṇṭha, Paramānanda, Mākhanalāla Trivedin, Rāma, Rāmadatta, Lakṣmīpati, Vināyaka, Śrīpati, and Sadāśiva. It was also published several times in the nineteenth century along with the commentaries of Gokulanātha, Divākara, and Viśvanātha.

II. Manuscripts of the *Makaranda:*

Śāstrī, Not. 1907 348 23 ff. Saṃ. 1815 = A.D. 1758.
Bombay U Desai 1376 38 ff. Saṃ. 1816 = 1759.
Vārāṇasī 34345 ff. 1–77. Saṃ. 1868 = A.D. 1811.
Mithila 244A 8 ff. Śaka 1750 = A.D. 1828.
Calcutta Sanskrit Coll. 89 32 ff. Saṃ. 1904 = A.D. 1847.
Jammu and Kashmir 2756G 28 ff. Saṃ. 1909 = A.D. 1852.
Mithila 137 5 ff. Śaka 1777 = A.D. 1855.
Mithila 244 9 ff. Śaka 1782 = A.D. 1860.
Mithila 406 6 ff. Śaka 1795, San. 1280 = A.D. 1873.
Mithila 24 3 ff. Śaka 1801 = A.D. 1879.
Mithila 244B 4 ff. Śaka 1801 = A.D. 1879
ABSP 1113 ff. 1–29.
Alwar 1891
Anup 4948 42 ff.
Anup 4949 44 ff.
Anup 4950 22 ff.
Anup 4951 14 ff.
Anup 4952 9 ff.
Anup 4953 6 ff.
Anup 4954 4 ff.
AS Bengal 6875 (G2647) 44 ff.
Baroda 3225 39 ff.
Baroda 3227 34 ff. (with ṭīkā of Viśvanātha).
Bharatpur 19
Bharatpur 21
BM 474 (Add. 26,488) I 10 ff. (with ṭīkā of Harikarṇa)
Bombay U 358 27 ff.
BORI 542 of 1875/76 12 ff. (from Delhi)
BORI 446 of 1895/98 29 ff.
Calcutta Sanskrit Coll. 53 26 ff.
CP, Hiralal 3736 (from Nagpur)
CP, Hiralal 3737 (from Maṇḍlā)
CP, Hiralal 3738 (from Valgaon in Amraotī District)
CP, Hiralal 3739 (from Haṭṭā in Damoh District)
IO 2954 (2476a)
Jammu and Kashmir 3072 3 ff.
Jammu and Kashmir 3166 12 ff.
Kathmandu 297 (III 471) 39 ff.
Kathmandu 298 (I 1191) 23 ff.
Kavīndrācārya 895
Kotah 188 3 pp.
LDI 7230 23 ff.
Oudh (1873) 16 10 pp.
*Poleman 4853, 4857, 4862 (Smith Indic 166) 31 ff.
*Poleman 4877 (Smith Indic 185) ff. 1-39v.
*Poleman 4881 (U Penn 1837) 20 ff.
*Poleman 4882 (U Penn 1788) ff. 1–25a and 25b–38.
*Poleman 4945 (Smith Indic 194) f. 12v.
*Poleman 4946 (Smith Indic MB) XXXVI. 1 f.
*Poleman 5188 (Smith Indic 129 B) ff. 2–21.
SOI 3388 53 ff. (with ṭīkā of Divākara).
SOI 7903/3
SOI 9429
SOI 9432
Vārāṇasī 34344. ff. 1–6 and 14.
Vārāṇasī 34639. ff. 1–30.
Vārāṇasī 34640. ff. 1–47.
Vārāṇasī 34641. 1 f.
Vārāṇasī 34751. 1 f.
Vārāṇasī 34984. ff. 1–2, 2–33, 1–10, 1–2, and 2–9.
Vārāṇasī 35090. ff. 1–35.
Vārāṇasī 35421. ff. 1–15 and 15–25 (with ṭīkā of Vināyaka). Identical with Benares (1903) 1310 25 ff. (with ṭīkā).
Vārāṇasī 35584. ff. 1–21. Identical with Benares (1888) 75 21 ff. Identical with Benares (1869) XV 3 21 ff.
Vārāṇasī 35585. ff. 1–13.
Vārāṇasī 35699. ff. 1–19.
Vārāṇasī 35712. ff. 3–4.
Vārāṇasī 35713. ff. 2–3.
Vārāṇasī 35714. ff. 1–42.
Vārāṇasī 35884. ff. 1–43.
Vārāṇasī 35885 ff. 1–24. Identical with Benares (1903) 1291 24 ff.
Vārāṇasī 36019. ff. 7–10 and 28–30.
Vārāṇasī 36085. ff. 3–10 and 14.
Vārāṇasī 36178. 11 ff.
Vārāṇasī 36824. ff. 1–22 and 24.
Vārāṇasī 37119. ff. 1–6.
Vārāṇasī 37120. ff. 1–20 (with ṭīkā of Divākara).
VVRI 2652 15 ff. (with ṭīkā of Viśvanātha).

III. Tables.

A. Tithis. (Tithisaurabhāṇi)

The mean tithis are tabulated for 1 to 16 single years (table 2) and for periods of 16 collected years (table 1). As in the table for single years integer numbers of tithis are recorded, the parameters for the columns dealing with days differ in the two tables; in table 1 that parameter is 1;15,23,15 days; in table 2 it is 1;11,41,40 days. As the tithi-parameter for table 1 is 11;3,45 tithis per year, this means that 6,11;3,45 tithis equal 6,5;15, 23,15 days, or 1 tithi equals 0;59,3,49 days. In table 2, integer numbers of tithis (11) are added for each year from 1 to 15, and then an extra tithi (12 in all) is entered under year 16 to make the yearly parameter 11;3,45. Therefore, in table 2, 6,11 tithis equal 6,5;11, 41,40 days, or 1 tithi equals 0;59,3,40 days.

A similar phenomenon occurs in the last column, where the parameter for 1 to 15 years is 15;12,35,48 ghaṭikās, but the entry for year 16 is so increased that the yearly parameter becomes 15;20,38,33,45 ghaṭikās. The same situation holds with regard to this column in the nakṣatra and yoga tables, but its function, as that of the whole column, is not yet understood.

1. Table of tithis (epact) for Śaka 1400 to 1976 (A.D. 1478 to 2054) in 16-year intervals. There are 4 columns. Column 1 designates the year; column 2 the tithis (the parameter is 27, or 11;3,45 tithis per year); column 3 the week-day and fraction on which the solar year begins (the parameter is 6;6,12, or 1;15,23,15 days per year); and column 4 the vallī for the year (the parameter is 5;30,17, or 15;20,38,33,45 ghaṭikās per year).

Manuscripts: 4853 (Smith Indic 166 ff. 1–1v); 4877 (Smith Indic 185 f. 1); 4881 f. 1v; 4882 f. 1; 5188 (Smith Indic 129 B f. 2).

2. Table of tithis for 1 to 16 years, also in 4 columns. Column 1 indicates the year-number; column 2 indicates the (integer) tithi; column 3 gives, for years 1 to 15, the week-day and fraction on which 6,11 tithis of the previous year end, and, for year 16, that on which the solar year begins (the parameter for years 1 to 15 is 1;11,41,40 days); and column 4 gives the corresponding vallīs (the parameter for years 1 to 15 is 15;12,35,48 ghaṭikās).

Manuscripts: 4853 (Smith Indic 166 f. 2); 4877 (Smith Indic 185 f. 1 and f. 1v); 4881 f. 2; 4882 f. 1; 5188 (Smith Indic 129 B f. 2).

3. Table of days in the pakṣas of a year, in 5 columns. Column 1 indicates the argument, 0 to 27 pakṣas; column 2, the week-day and fraction on which the pakṣa begins; column 3, a negative cālana measured in ghaṭikās; column 4, the corresponding vallīs modulo 60; and column 5, a positive cālana measured in ghaṭikās.

Manuscripts: 4877 (Smith Indic 185 f. 2); 4881 f. 2v; 4882 f. 1v.

4. A table for determining variations in the length of a tithi. It is arranged for 0 to 59 entries (each corresponding to 6° of lunar anomalistic motion) horizontal, and for 0 to 59 entries (each corresponding to 0;6° of lunar anomalistic motion) vertical. The maximum correction to the tithikendra in the table is 49;54 ghaṭikās, the minimum 0;0, and the mean 24;57.

Manuscripts: 4853 (Smith Indic 166 ff. 3–5v); 4857 (Smith Indic 166 ff. 10–18v); 4877 (Smith Indic 185 ff. 2v–3v); 4881 ff. 3–3v; 4882 ff. 2–6v; 5188 (Smith Indic 129 B ff. 2v–4).

4a. Table of mean tithis (0;59,3 days) for 0 to 404 tithis. Also a column containing the kendra for each tithi measured in ghaṭikās; the interval between entries in this column is 2;8,30 ghaṭikās. This table is not a part of the *Makaranda*.

Manuscripts: 5188 (Smith Indic 129 B ff. 4v–7v).

B. Nakṣatras (Nakṣatrasaurabha).

5. Table of nakṣatras for Śaka 1400 to 1976 (A.D. 1478 to 2054) in 24-year intervals. There are 4 columns. Column 1 designates the year; column 2, the nakṣatras (the parameter is 23, or 9;57,30 nakṣatras per year); column 3 the week-day and fraction on which the nakṣatra indicated in column 2 begins (the parameter is 2;12,35, or 1;15,31,27,30 days per year); and column 4 the corresponding vallī (the parameter is 8:22,36, or 15;20,56,30 ghaṭikās per year).

The Moon travels through 351 nakṣatras in 13 sidereal months; so the parameters of columns 2 and 3 indicate that it travels 1,20,12;46,40° in 6,5;15,31,27,30 days, or 13;10,34,52° per day.

Manuscripts: 4857 (Smith Indic 166 f. 8); 4877 (Smith Indic 185 f. 6); 4881 f. 4; 4882 f. 7; 5188 (Smith Indic 129 B f. 8).

6. Table of nakṣatras for 1 to 24 years in 4 columns. Column 1 indicates the year-number; column 2 the (integer) nakṣatra (for 1 to 24 the parameter varies between 10 and 9); column 3, the week-day on which the nakṣatra begins (for 1 to 23 the parameter is 1;12, 16,20 days); and column 4 the corresponding vallī (for 1 to 23 the parameter is 14;42,20,22,30 ghaṭikās).

Manuscripts: 4857 (Smith Indic 166 f. 8); 4862 (Smith Indic 166 ff. 19–18v); 4877 (Smith Indic 185 f. 6); 4881 f. 4; 4882 f. 7; 5188 (Smith Indic 129 B f. 8).

7. Table of week-days with which sidereal months begin in 5 columns. Column 1 indicates the sidereal month from 0 to 14; column 2, the week-day and fraction with which it begins; column 3, a positive cālana measured in ghaṭikās; column 4, the corresponding vallīs tabulated modulo 60; and column 5, a positive cālana measured in ghaṭikās.

Columns 1 and 2 are:

sidereal month	week-day	sidereal month	week-day
1	6;20,10	8	1;33,0
2	5;40,18	9	0;51,28
3	4;59,49	10	0;10,47
4	4;19,10	11	6;30,24
5	3;37,44	12	5;50,24
6	2;56,3	13	5;10,33
7	2;14,37	14	4;30,45

Manuscripts: 4862 (Smith Indic 166 f. 19v); 4877 (Smith Indic 185 f. 4); 4881 f. 4v; 4882 f. 7v.

8. A table for determining the length of a nakṣatra, set

up as is table 4. The maximum entry is 56;8 ghaṭikās, the minimum 0;0, and the mean 23;4.

Manuscripts: 4862 (Smith Indic 166 ff. 20–31v); 4877 (Smith Indic 185 ff. 4v–5v); 4881 ff. 5–5v; 4882 ff. 8–12v; 5188 (Smith Indic 129 B ff. 8v–10).

C. Yogas. (Yogakṣepaka)

9. Table of yogas for Śaka 1400 to 1976 (A.D. 1478 to 2054) in 24-year intervals. There are 4 columns. Column 1 designates the year; column 2, the yogas; column 3, the week-day and fraction on which the yoga indicated in column 2 begins; and column 4, the corresponding vallī. The parameters are exactly the same as those for table 5, since at the end of 24 years both the Sun and the Moon are entering nakṣatras.

Manuscripts: 4862 (Smith Indic 166 f. 19); 4877 (Smith Indic 185 f. 6v); 4881 f. 6; 4882 f. 13; 5188 (Smith Indic 129 B f. 14).

10. Table of yogas for 1 to 24 years in 4 columns. Column 1 indicates the year-number; column 2 is identical with column 2 of table 6; column 3 indicates the week-day (for 1 to 23, the parameter is 1;12,16,45 days per year); and column 4 the vallī (for 1 to 23, the parameter is 14;42,21,20 ghaṭikās).

Manuscripts: 4877 (Smith Indic 185 f. 6v); 4881 f. 6; 4882 f. 13; 5188 (Smith Indic 129 B f. 14).

11. Table of week-days related to yogas in 5 columns. Column 1 indicates the argument from 0 to 15; column 2, the week-day and fraction; column 3, a negative cālana measured in ghaṭikās; column 4, the corresponding vallīs tabulated modulo 60; and column 5, a positive cālana measured in ghaṭikās.

Manuscripts: 4877 (Smith Indic 185 f. 7); 4881 f. 6v; 4882 f. 13v.

12. A table for determining the length of a yoga, set up as is table 4. The maximum entry is 42;54 ghaṭikās, the minimum 0;0, and the mean 21;27.

Manuscripts: 4877 (Smith Indic 185 ff. 7v–8v); 4881 ff. 7–7v; 4882 ff. 14–18v; 5188 (Smith Indic 129 B ff. 14v–16).

D. Saṅkrāntis.

13. Table of week-days on which the Sun enters Aries after each period for 24 years from Śaka 1400 to 1976 (A.D. 1478 to 2054). The epoch position for Śaka 1400 is 6;25,41, i.e. Friday, 27 March 1478.

Manuscripts: 4877 (Smith Indic 185 f. 9); 4881 f. 8; 4882 f. 19; 5188 (Smith Indic 129 B f. 20).

14. Table of week-days on which the Sun enters Aries for 1 to 24 years. The parameter is 1;15,31,30 days per year.

Manuscripts: 4877 (Smith Indic 185 f. 9); 4881 f. 8; 4882 f. 19; 5188 (Smith Indic 129 B f. 20).

15. Table of week-days on which the Sun enters each zodiacal sign:

Aries	0;0,0	Libra	4;55,48
Taurus	2;57,1	Scorpio	6;48,31
Gemini	6;23,1	Sagittarius	1;17,11
Cancer	3;0,51	Capricorn	2;36,4
Leo	6;30,4	Aquarius	4;3,14
Virgo	2;30,29	Pisces	5;53,21

Manuscripts: 4857 (Smith Indic 166 f. 9); 4877 (Smith Indic 185 f. 9); 4881 f. 8; 4882 f. 19; 5188 (Smith Indic 129 B f. 20v).

16. Table of week-days on which the Sun enters each of the 27 nakṣatras.

Aśvinī	0;0,0	Svāti	4;36,45
Bharaṇī	6;41,24	Viśākhā	3;56,46
Kṛttikā	6;30,0	Anurādhā	3;5,47
Rohiṇī	6;24,35	Jyeṣṭhā	2;12,47
Mṛgaśiras	6;23,36	Mūla	1;17,11
Ārdrā	6;24,27	Pūrvāṣāḍhā	0;29,0
Punarvasu	6;29,38	Uttarāṣāḍhā	6;19,50
Puṣya	6;32,29	Śravaṇa	5;23,0
Āśleṣā	6;32,40	Dhaniṣṭhā	4;30,0
Maghā	6;30,4	Śatabhiṣak	3;40,0
Pūrvaphālgunī	6;19,41	Pūrvabhādrapadā	2;51,0
Uttaraphālgunī	6;5,42	Uttarabhādrapadā	2;11,0
Hasta	5;43,43	Revatī	1;42,0
Citra	5;11,44		

Manuscripts: 4857 (Smith Indic 166 ff. 9–9v); 4877 (Smith Indic 185 f. 9); 4881 f. 8; 4882 f. 19.

The next three tables are not from the *Makaranda*.

16a. Table of precession for 20-yr. periods from Śaka 1680 to 1760 (A.D. 1758 to 1838), and for 1 to 20 years. The precession for Śaka 1680 amounts to 18;53,6°, and the yearly parameter is 0;0,54°. Precession, then, is taken to have been 0 in Śaka 421 (A.D. 499), the date at which Āryabhaṭa I wrote his *Āryabhaṭīya*. The parameter 0;0,54° per year is equal to the common Arabic and Indian value of 1° every 66 years.

Manuscripts: 4881 f. 8v.

16b. Table of week-days on which the Sun enters each of the 12 signs of the zodiac in a tropical year.

Aries	1;17,4	Libra	5;4,0
Taurus	2;39,20	Scorpio	0;15,0
Gemini	5;50,0	Sagittarius	1;55,0
Cancer	2:24,0	Capricorn	3;17,0
Leo	5;55,0	Aquarius	4;37,0
Virgo	2;17,0	Pisces	6;12,0

Manuscripts: 4881 f. 8v.

16c. Table of week-days on which the Sun enters each of the 28 nakṣatras.

Aśvinī	0;0,0	Svāti	4;36,6
Bharaṇī	6;41,2	Viśākhā	3;53,9
Kṛttikā	6;30,5	Anurādhā	3;5,12
Rohiṇī	6;24,7	Jyeṣṭhā	2;17,15
Mṛgaśiras	6;21,10	Mūla	1;17,18
Ārdrā	6;24,12	Pūrvāṣāḍhā	0;21,16

Punarvasu	6;29,15	Uttarāṣāḍhā	6;24,13
Puṣya	6;32,13	Abhijit	2;8,58
Āśleṣā	6;32,10	Śravaṇa	5;26,11
Maghā	6;30,4	Dhaniṣṭhā	4;30,8
Pūrvaphālgunī	6;19,5	Śatabhiṣak	3;40,6
Uttaraphālgunī	6;5,3	Pūrvabhādrapadā	2;51,3
Hasta	5;43,0	Uttarabhādrapadā	2;11,1
Citra	5;11,3	Revatī	1;42,11

Manuscripts: 4945 f. 12v; 5188 (Smith Indic 129 B ff. 20 and 21).

17. Table of vallīs for the saṅkrāntis in 3 parts:

Part 1. Vallīs for 57-year periods from Śaka 1400 to 1799 (A.D. 1478 to 1877). The epoch position for Śaka 1400 is 0;7,44,34,54.

Part 2. Vallīs for 1 to 57 years: 57 years contain three 19-year cycles, so there are 21 intercalary months (adhimāsas).

Part 3. Vallīs for 1 to 24 pakṣas (12 synodic months; a normal year) and for 2 additional pakṣas (the adhimāsa, to be added for an intercalary year). The parameter for a normal year is 0;0,0,5,54; for an intercalary year, 0;0,0,6,24. The year begins with Caitrāmāvāsya—i.e., new moon of the month Caitra; but the months begin with a full moon.

Manuscripts: 4877 (Smith Indic 185 ff. 10v–11); 4881 f. 9; 4882 f. 19v.

E. Mean motion (vāṭikā)

These tables are set up for each planet so that the intervals of the argument (k = 1 to 1,0) can be taken as 10 ghaṭikās, 10 days, or any other multiple of 10 ghaṭikās by 60 raised to any power. The entry for k = 6, therefore, can be read as the mean daily motion, and we can construct the following table:

Sun	0;59,8,10,10,24,12,30°
Moon	13;10,34,52,3,49,4,12
Lunar Anomaly	13;3,53,53,33,7,40,12
Mars	0;31,26,28,11,8,56,30
Mercury (anomaly)	56;53,35,49,28,32,56,6
(1,0—Mercury	3;6,24,10,31,27,3,54)
Jupiter	0;4,59,8,48,35,47,36
Venus (anomaly)	59;23,0,26,33,7,19,36
(1,0—Venus	0;36,59,33,26,52,40,24)
Saturn	0;2,0,22,53,25,46,48
Lunar Node	59;56,49,15,5,0,16,36
(Lunar Node—1,0	−0;3,10,44,54,59,43,24)

18. Mean motion table of the Sun.

Manuscripts: 4877 (Smith Indic 185 ff. 11v–12); 4881 f. 9v; 4882 f. 20.

19. Mean motion table of the Moon.

Manuscripts: 4877 (Smith Indic 185 ff. 12–12v); 4881 f. 10; 4882 f. 20v.

20. Mean motion table of the Lunar Anomaly.

Manuscripts: 4877 (Smith Indic 185 ff. 12v–13v); 4881 f. 10v; 4882 f. 21.

21. Mean motion table of Mars.

Manuscripts: 4877 (Smith Indic 185 ff. 13v–14); 4881 f. 11; 4882 f. 21v.

22. Mean motion table of Mercury's anomaly.

Manuscripts: 4877 (Smith Indic 185 ff. 14v–15 and ff. 38–38v); 4881 f. 11v; 4882 f. 22.

23. Mean motion table of Jupiter.

Manuscripts: 4877 (Smith Indic 185 ff. 15–16); 4881 f. 12; 4882 f. 22v.

24. Mean motion table of Venus' anomaly.

Manuscripts: 4877 (Smith Indic 185 ff. 16–16v and ff. 38v–39v); 4881 f. 12v; 4882 f. 23.

25. Mean motion table of Saturn.

Manuscripts: 4877 (Smith Indic 185 ff. 16v–17); 4881 f. 13; 4882 f. 23v.

26. Mean motion table of the Lunar Node.

Manuscripts: 4877 (Smith Indic 185 ff. 17–18); 4881 f. 13v; 4882 f. 24.

The next two tables are probably not from the *Makaranda.*

26a. Table of mean daily motions of the planets in minutes and seconds.

Sun	59,8	Venus	96,8
Moon	790,35	Saturn	2,0
Mars	31,26	Lunar Apogee	6,41
Mercury	245,32	Lunar Node	−3,11
Jupiter	5.0		

Cf. table 24 of the *Grahaprabodhasāriṇī.*

Manuscripts: 4877 (Smith Indic 185 f. 18).

26b. Table of solar daily motion in each of the zodiacal signs.

Aries	Leo	0;57,33	Sagittarius	1;1,18
Taurus	Virgo	0;58,34	Capricorn	1;1,12
Gemini	Libra	0;59,42	Aquarius	1;0,15
Cancer	Scorpio	1;0,52	Pisces	0;59,18

Manuscripts: 4877 (Smith Indic 185 f. 18).

F. Equations.

27. Table of the equation of the center for the Sun in 2 parts. Part 1 tabulates the equation for 0° to 90° of argument; the maximum equation is 2;10,32°, and the solar apogee is at Gemini 17;16,51°. Part 2 gives the increment to be added to the mean daily motion (or the decrease to be subtracted from it) for 0° to 90° of argument; the maximum increment is 0;2,18°.

Manuscripts: 4877 (Smith Indic 185 ff. 18v–19); 4881 f. 14; 4882 f. 24v.

28. Table of the equation of the center for the Moon for 0° to 90°; both the equation and the increment are tabulated. The maximum equation is 5;2,48°; the maximum increment 1;9,39°.

Manuscripts: 4877 (Smith Indic 185 ff. 19v–20v); 4881 f. 14v; 4882 f. 25a.

29. Table of equations for Mars for 0° to 180°, in 4 columns. Column 1 indicates the argument; column 2, the equation of the conjunction; column 3, the equation of the center; and column 4, the hypotenuse (karṇa). The maximum equation of the conjunction is 40;16° at 130° to 132° of argument; the maximum equation of the center is 11;32° at 93° to 98° of argument; and the maximum hypotenuse is 99 at 0° to 11° of argument.

Also given are the following parameters:

apogee at Leo 10;2,28°
first station at elongation of 164°
second station at elongation of 196°
first visibility at elongation of 28°
last visibility at elongation of 332°
node at Taurus 10;3,28°

Manuscripts: 4877 (Smith Indic 185 ff. 20v–22v;) 4881 ff. 15–15v; 4882 ff. 25b–26.

30. Table of equations for Mercury for 0° to 180°, set up as is table 29. The maximum equation of the conjunction is 21;31° at 109° to 113° of argument; the maximum equation of the center is 4;28° at 92° to 93° of argument; and the maximum hypotenuse is 82 at 0° to 16° of argument.

Also given are the following parameters:

apogee at Scorpio 10;27,45°
first station at anomaly of 144°
second station at anomaly of 216°
last visibility in the West at anomaly of 155°
first visibility in the East at anomaly of 205°
last visibility in the East at anomaly of 310°
first visibility in the West at anomaly of 50°
node at Aries 20;41,24°.

Manuscripts: 4877 (Smith Indic 185 ff. 23–25v); 4881 ff. 16–16v; 4882 ff. 26v–27v.

31. Table of equations for Jupiter for 0° to 180°, set up as is table 29. The maximum equation of the conjunction is 11;31° at 100° to 101° of argument; the maximum equation of the center is 5;6° at 91° to 95° of argument; and the maximum hypotenuse is 72 at 0° to 11° of argument.

Also given are the following parameters:

apogee at Virgo 21;20,48°
first station at elongation of 130°
second station at elongation of 230°
last visibility at elongation of 346°
first visibility at elongation of 14°
node at Gemini 2;0,0°.

Manuscripts: 4877 (Smith Indic 185 ff. 25v–28); 4881 ff. 17–17v; 4882 ff. 28–29.

32. Table of equations for Venus for 0° to 180°, set up as is table 29. The maximum equation of the conjunction is 46;24° at 136° to 138° of argument; the maximum equation of the center is 1;45° at 85° to 98° of argument; and the maximum hypotenuse is 104 at 0° to 5° of argument.

Also given are the following parameters:

apogee at Gemini 19;51,23°
first station at anomaly of 167°
second station at anomaly of 193°
last visibility in the West at anomaly of 177°
first visibility in the East at anomaly of 183°
last visibility in the East at anomaly of 336°
first visibility in the West at anomaly of 24°
node at Taurus 19;40,43°

Manuscripts: 4877 (Smith Indic 185 ff. 28–30v); 4881 ff. 18–18v; 4882 ff. 29v–30v.

33. Table of equations for Saturn for 0° to 180°, set up as is table 29. The maximum equation of the conjunction is 6;22° at 94° to 99° of argument; the maximum equation of the center is 7;40° at 92° to 96° of argument; and the maximum hypotenuse is 66 at 0° to 14° of argument.

Also given are the following parameters:

apogee at Scorpio 26;37,34°
first station at elongation of 113°
second station at elongation of 247°
last visibility at elongation of 343°
first visibility at elongation of 17°
node at Cancer 10;21,40°

Manuscripts: 4877 (Smith Indic 185 ff. 30v–32); 4881 ff. 19–19v; 4882 ff. 31–32.

G. Miscellaneous. (not from the *Makaranda*)

34. Table of nakṣatracaraṇas, each of which is 1/4 of a nakṣatra or 3;20°; they are, therefore, equal to navāṃśas.

Manuscripts: 4882 ff. 32–33.

35. Mean motion tables (kṣepa) for Sun, Moon, Lunar Apogee, Mars, Mercury, Jupiter, Venus, Saturn, Rāhu, and Ketu for 32, 31, 30, 29, 8, and 7 days. The mean daily motions are:

Sun	0;59,8,10°	Jupiter	0;4,59,8°
Moon	13;10,34,52	Venus (conj)	1;36,7,42
Lunar Apogee	0;6,40,58	Saturn	0;2,0,22
Mars	0;31,26,28	Rāhu	–0;3,10,46
Mercury (conj)	4;5,32,20	Ketu	–0;3,10,46

These are not the mean daily motions of the *Makaranda*.

Manuscripts: 4882 ff. 33–33v.

36. Mean motion tables (cālika) of Sun, Moon, Lunar Apogee, Mars, Mercury (conj.), Jupiter, Venus (conj.), Saturn, and the Lunar Node for 1, 13, 14, 15, 16, and 17 days, and for 354, 355, 384, and 385 days. The mean daily motions are:

Sun	0;59,8,10,10,24,12,30°
Moon	13;10,34,52,3,49,4,12
Lunar Apogee	0;6,40,56,30,41,24,8
Mars	0;31,26,28,11,8,56,34

Mercury (conj)	4;5,32,20,41,51,16,25
Jupiter	0;4,59,8,48,35,47,37
Venus (conj)	1;36,7,43,37,16,52,54
Saturn	0;2,0,22,53,25,46,48
Lunar Node	-0;3,10,44,54,59,43,25

These parameters are identical with those of *Makaranda.*

Manuscripts: 4877 (Smith Indic 185 ff. 36–37v); 4881 f. 20; 4882 ff. 33v–34.

37. Tables of Greek-letter phenomena.

a: first and second stations.

b: last visibility in the West and first visibility in the East.

c: first visibility in the West and last visibility in the East.

The parameters are identical with those of the *Makaranda.*

Manuscripts: 4857 (Smith Indic 166 f. 9v); 4877 (Smith Indic 185 f. 33v); 4882 f. 34; 4946 XXXVI f. 1r.

38. A Dinamānapattra; for details see the descriptions of the manuscripts.

Manuscripts: 4877 (Smith Indic 185 ff. 9v–10); 4882 f. 35.

38a. Dinamānapattras for two localities; the argument is expressed in units equal to 6° of solar motion. The maximum lengths of daylight, at 15 (= 90°), are 34;5 ghaṭikās and 34;18 ghaṭikās. At the top of the table it is stated that precession equals 17;16°; at a rate of 1° every 66 years, this dates the table to the middle of the seventeenth century.

Manuscripts: 4877 (Smith Indic 185 ff. 32v–33).

39. Table of week-days on which the Sun enters the zodiacal signs taken from a Persian Source. Its title is: atha phārasī (فارسى) saṃkrāṃtakṣepakakoṣṭhaka.

hamala	(حمل)	0;0,0
śaura	ثور	2;37,42
jaujā	جوزاء	5;43,50
saratā	(سرطان)	2;7,20
asada	(اسد)	5;32,55
śuṃbula	(سنبلة)	1;45,13
mejā	(ميزان)	4;31,51
akra	(عقرب)	6;41,10
kauśa	(قوس)	1;33,38
hadī	حدى	3;1,53
dalaḥ	دلو	4;27,55
hūta	حوت	6;8,2

Manuscripts: 4882 f. 35.

40. Table of week-days with which years at 30-year intervals from Śaka 1700 to 1790 (A.D. 1778 to 1868) begin, and with which 1 to 30 years begin. The yearly parameter is 1;14,33,8 days, so that a tropical year is 6,5;14,33,8 days long. The epoch position for Śaka 1700 is 6;33,7.

This table also is derived from the Persian source; its title is: atha tahavīla (تحويل) hamalādi phāraṣī sa⟨ṃ⟩krāṃti śākā.

Manuscripts: 4882 f. 35v.

41. A table of the ends of the nakṣatracaraṇas, and a table of their beginnings.

Manuscripts: 4882 ff. 36–36v. *Cf.* 4877 (Smith Indic 185 ff. 34–35).

42. Table of the duration of lunar eclipses (stithyardha) measured in ghaṭikās for 1 to 36 units of 0;3° (viṃśopakas). Since 1 digit equals 0;5°, a viṃśopaka equals 0;36 digits and 36 viṃśopakas equals 21;36 digits.

1	0;52	13	3;32	25	4;23
2	1;30	14	3;37	26	4;25
3	1;51	15	3;43	27	4;27
4	2;10	16	3;49	28	4;29
5	2;25	17	3;54	29	4;31
6	2;37	18	3;59	30	4;33
7	2;47	19	4;4	31	4;35
8	2;54	20	4;8	32	4;37
9	3;3	21	4;11	33	4;38
10	3;11	22	4;14	34	4;39
11	3;19	23	4;17	35	4;40
12	3;26	24	4;20	36	4;40

Manuscripts: 4882 f. 37.

43. Table of parallax of longitude (lambana) measured in palas (60 palas equal 1 ghaṭikā) for 1 to 17 units of 6° in the Moon's zenith distance.

1	43	10	238
2	84	11	240
3	118	12	239
4	151	13	235
5	176	14	229
6	197	15	221
7	214	16	208
8	226	17	180
9	234		

Manuscripts: 4882 f. 37v.

44. Table of solar altitude (unnati) measured in palas for 0 to 36 units of 10° (daśaka).

1	173,34	13	175,9	25	115,48
2	176,6	14	172,0	26	116,45
3	178,0	15	168,12	27	118,40
4	179,0	16	163,18	28	122,27
5	179,33	17	157,40	29	127,48
6	179,35 (*sic*)	18	152,48	30	133,51
7	179,52	19	144,24	31	140,30
8	179,51	20	137,42	32	147,27
9	179,42	21	131,3	33	154,9
10	179,42 (*sic*)	22	125,9	34	160,12
11	178,30	23	120,45	35	165,36
12	177,15	24	117,15	36	170,6

Manuscripts: 4882 f. 37v.

45. Table of parallax of latitude (avanati) measured in palas for 0 to 35 daśakas.

0	281,6	12	308,39	24	203,24
1	292,24	13	298,30	25	202,12
2	302,51	14	287,21	26	202,38
3	312,42	15	275,51	27	204,42
4	321,0	16	264,18	28	208,3
5	327,0	17	252,57	29	213,15
6	331,45	18	243,27	30	219,40
7	333,36	19	232,30	31	227,33
8	333,3	20	224,3	32	236,48
9	328,51	21	216,33	33	247,0
10	324,45	22	211,29	34	257,57
11	317,21	23	206,36	35	268,27

Manuscripts: 4882 f. 38.

46. Table of vikṣepas for the parallax of latitude for daśakas 34, 35, 0, 1, 2, 16, 17, 18, 19, and 20.

34	680,0	16	0,0
35	512,45	17	167,15
0	340,0	18	340,0
1	167,15	19	512,45
2	0,0	20	680,0

Manuscripts: 4882 f. 38v.

47. Table of half-duration of the eclipse (sthityardha) for 1 to 24 viṃśopakas.

1	0,39	9	2,16	17	2,41
2	1,18	10	2,21	18	2,42
3	1,26	11	2,25	19	2,43
4	1,40	12	2,28	20	2,44
5	1,48	13	2,32	21	2,44
6	1,57	14	2,35	22	2,45
7	2,4	15	2,37	23	2,45
8	2,10	16	2,39	24	0,0

Manuscripts: 4882 f. 38v.

48. Table of the apparent diameters measured in digits of the Moon and of the cone of the earth's shadow; the argument is the time it takes the Moon to travel one nakṣatra or 13;20° (nakṣatrabhoga). *Cf.* table 6 of the *Karaṇakesarī*. The argument increases from 0;56 days to 1;6 days as the daily progress of the Moon decreases from 14;17,8° to 12;7,17°. The maximum diameters of the Moon and the cone of the earth's shadow are respectively 11;34 and 29;34 digits, their minimum diameters 9:49 and 23;48 digits.

Manuscripts: 4877 (Smith Indic 185 f. 33).

49. Table of unknown purpose. The argument increases from 1 (= 31) to 25 (= 55), and the entries from 9,24 to 46,21.

Manuscripts: 4877 (Smith Indic 185 f. 33v).

50. Table entitled saṃkrānti, indicating the number of the day (modulo 30) on which the Sun enters each zodiacal sign.

Aries	10;46,0,22	Libra	10;57,0,14
Taurus	10;35,0,31	Scorpio	11;8,0,15
Gemini	10;27,0,37	Sagittarius	11;14,0,1
Cancer	10;26,0,37	Capricorn	11;15,0,0
Leo	10;33,0,31	Aquarius	11;8,0,5
Virgo	10;44,0,23	Pisces	10;58,0,13

Manuscripts: 4877 (Smith Indic 185 f. 33v).

51. Table of the duration of lunar eclipses measured in ghaṭikās. The argument increases from 1 to 22 digits, the entries from 1;29 to 4;39 ghaṭikās. *Cf.* table 42.

Manuscripts: 4877 (Smith Indic 185 f. 33v).

52. Double-entry table of unknown purport. The arguments, both horizontal and vertical, increase from 1 to 12 (zodiacal signs?); the entries range from 53 at 1, 1 to 101 at 6 (vertical), 9 (horizontal).

Manuscripts: 4853 (Smith Indic 166 f. 7v); 4877 (Smith Indic 185 f. 35v).

ANONYMOUS OF 1520

I. Manuscripts:

*Poleman 4798, 4799 (U Penn 1913 (a) ff.1–2v).
*Poleman 4896 (U Penn 705) 12 ff.
*Poleman 5108 (U Penn 1824) ff. 2–10.
*Poleman 5172 (U Penn 1877) 2 ff.

II. Tables:

A. Mean motions:

These tables are of two sorts: (a) of mean daily motions for 0 to 60 days, and for 1,0 to 1,10,0 days in steps of 1,0 days; and (b) for periods of 11 "years" (*cf.* the *Grahaprabodhasāriṇī*) for periods 17 to 33. Accompanying the (b) tables are the dhruvas, which are equal to 360° diminished by the motion of the planet in an 11-"year" period, and the kṣepakas, which are the epoch positions. These kṣepakas indicate that the epoch of the system was 18 March 1520.

	sidereal	tropical	sidereal-tropical
Saturn	285;21°	302°	−17°
Jupiter	212;16	232	−20

Mars	307;8	337	−30
Sun	349;41	3	−13
Venus' anomaly	230;9	(Venus) 321	
Mercury's anomaly	269;33	(Mercury) 340	
Moon	349;6	0	−11
Lunar apogee	317;33		
Lunar node	27;38		

Since our manuscripts use the periods 17 to 33, their authors or scribes were interested in the year 1707 to 1850 A.D.

The mean daily motions of the planets, from table (a), and the *dhruvas*, from table (b), are:

	mean daily motions	*dhruvas*
Saturn	0;2,0,23,4,37°	225;42°
Jupiter	0;4,59,8,34,18	26;18
Mars	0;31,26,31,3,35	55;32
Sun	0;59,8,10,17,9	1;49,11
Venus (anom.)	0;36,59,40,6,37	44;2
Mercury (anom.)	3;6,24,8,7,13	123;27
Moon	13;10,34,51,55,59	3;46,11
Lunar apogee	0;6,40,51,25,42	272;45
Lunar node	−0;3,10,48,25,15	212;50

The solar parameter gives a year-length of 6,5;15,30, 49,43 days.

Manuscripts:

1. Sun, table (a)	4896 ff. 1–1v; 5108 ff. 2–2v.
2. Sun, table (b)	4896 f. 1v; 5108 f. 2v.
3. Moon, table (a)	4896 ff. 2–2v; 5108 ff. 3–3v.
4. Moon, table (b)	4896 f. 2v; 5108 f. 3v.
5. Lunar apogee, table (a)	4896 ff. 3–3v; 5108 ff. 4–4v.
6. Lunar apogee, table (b)	4896 f. 3v; 5108 f. 4v.
7. Lunar node, table (a)	4896 ff. 4–4v; 5108 ff. 10–10v.
8. Lunar node, table (b)	4896 f. 4v; 5108 f. 10v.
9. Mars, table (a)	4896 ff. 5–5v; 5108 ff. 5–5v.
10. Mars, table (b)	5108 f. 5v
11. Mercury's anomaly, table (a)	4896 ff. 6–6v; 5108 ff. 6–6v.
12. Mercury's anomaly, table (b)	4896 f. 6v; 5108 f. 6v.
13. Jupiter, table (a)	4896 ff. 7–7v; 5108 ff. 7–7v.
14. Jupiter, table (b)	4896 f. 7v; 5108 f. 7v.
15. Venus' anomaly, table (a)	4896 ff. 8–8v; 5108 ff. 8–8v.
16. Venus' anomaly, table (b)	4896 f. 8v; 5108 f. 8v.
17. Saturn, table (a)	4896 ff. 9–9v; 5108 ff. 9–9v.
18. Saturn, table (b)	4896 f. 9v; 5108 f. 9v.

B. Equations of the Sun and Moon

19. Equation of the center for the Sun. Table for 0° to 90° in 4 columns. Column 1 lists the degrees; column 2 the equations; column 3 the differences between entries in column 2; and column 4 the increment to be added to (or decrease to be subtracted from) the mean daily motion of the Sun to get its true daily motion. The maximum equation is 2;10,45°; the maximum increment 0;2,16°. Manuscripts: 4798 (U Penn 1913 (a) ff. 1–1v); 4896 ff. 10–10v; 5172 ff. 1–1v.

20. Equation of the center for the Moon. This table is set up as is table 19. The maximum equation is 5;1,40°; the maximum increment 1;8,1°. Manuscripts: 4799 (U Penn 1913 (a) ff. 2–2v); 4896 ff. 11–11v; 5172 ff.2–2v.

THE *TITHICINTĀMAṆI* OF GAṆEŚA

I. The Life of Gaṇeśa:

Gaṇeśa, the son of Keśava and Lakṣmī, was born at Nandigrāma (Nandod near the mouth of the Narmadā) in Śaka 1429 (A.D. 1507). A catalogue of his numerous works on jyotiḥśāstra and dharmaśāstra is given by his nephew Nṛsiṃha in the following two verses from his commentary on the *Grahalāghava:*

kṛtvādau grahalāghavaṃ laghubṛhattithyādicintā-
maṇiṃ
satsiddhāntaśiromaṇeś ca vivṛtiṃ līlāvatīvyākṛtim /
śrīvṛndāvanaṭīkikāṃ ca vivṛtiṃ mauhūrtatattvasya
vai
sacchrāddhādivinirṇayaṃ suvivṛtiṃ chandorṇavā-
khyasya vai / /
sudhīrañjanaṃ tarjanīyantrakaṃ ca sukṛṣṇāṣṭamīnirṇa-
yaṃ holikāyāḥ /
laghūpāyayātāṃs tathānyān apūrvān gaṇeśo gurur
brahmanirvāṇam āpat / /

Of his surviving works on jyotiḥśāstra, we know that he wrote the *Grahalāghava* in Śaka 1442 (A.D. 1520); the *Pātasāraṇī* in Śaka 1444 (A.D. 1522); the *Tithicintāmaṇi* in Śaka 1447 (A.D. 1525); the *Buddhivilāsinī,* a commentary on Bhāskara II's *Līlāvatī*, in Śaka 1467 (A.D. 1545); the *Bṛhattithicintāmaṇi* in Śaka 1474 (A.D. 1552); and the *Vivāhadīpikā*, a commentary on Keśavārka's *Vivāhavṛndāvana,* in Śaka 1476 (A.D. 1554). The date of his commentary on his father's *Muhūrtatattva* is unknown; but it is referred to in his *Vivāhadīpikā* and therefore antedates 1554. For Gaṇeśa's works on dharmaśāstra see P. V. Kane, *History of Dharmaśāstra*, vol. 1, p. 692.

The *Tithicintāmaṇi* which we are here concerned with has several times been published in India. There are commentaries by Gaṇeśa's nephew, Nṛsiṃha (b. 1586), by Viśvanātha (1634) of Benares, and by Bāpū and Yajñeśvara. That by Viśvanātha has been published.

II. Manuscripts of the *Tithicintāmaṇi:*

PL, Buhler 422 5 ff. Saṃ. 1626 = A.D. 1569.
Vārāṇasī 35336 ff. 1–38. Saṃ. 1685 = A.D. 1628.
Vārāṇasī 36648 ff. 1–5. Saṃ. 1724 = A.D. 1667.
*Poleman 4679, 4678, 4681, 4682 (Smith Indic 94) 40 ff. Śaka 1652 = A.D. 1730.

*Poleman 4906 (U Penn 1891) 15 ff. Śaka 1683 = A.D. 1761.
*Poleman 4680 (Smith Indic 92) ff. 1–4v. Śaka 1704 = A.D. 1782.
Vārāṇasī 35329 ff. 1–7. Saṃ. 1843 = A.D. 1786.
DC 396 7 ff. Śaka 1712 = A.D. 1790.
Nagpur 818 (1017) 6 ff. Saṃ. 1848 = A.D. 1791.
VVRI 5172 31 ff. Śaka 1718 = A.D. 1796.
*Poleman 4814 (U Penn 1859) 12 ff. Saṃ. 1854 = A.D. 1797.
Vārāṇasī 35754 ff. 1–20. Śaka 1726 = A.D. 1804.
Vārāṇasī 35333 ff. 1–22. Saṃ. 1864, Śaka 1729 = A.D. 1807. Identical with Benares (1903) 1163 22 ff. Saṃ. 1864 = A.D. 1807.
Gujarāt Vidyā Sabhā 2547 ff. 1–4. Śaka 1739 = A.D. 1817. (*Pañcāṅgasādhana*).
AS Bombay 240 11 ff. Śaka 1745 = A.D. 1823.
SOI 2091 3 ff. Śaka 1750 = A.D. 1828.
BM 463 (Add. 14,365 e) 4 ff. A.D. 1829.
BORI 420 of A 1881/82 6 ff. Śaka 1751 = A.D. 1829.
Vārāṇasī 34763 ff. 1–14, 1–6, and 6–23. Saṃ. 1887 = A.D. 1830 (with ṭīkā).
PL, Buhler 182 3 ff. Saṃ. 1904 = A.D. 1848.
Jammu and Kashmir 3082 3 ff. Saṃ. 1911 = A.D. 1854.
Vārāṇasī 35486. ff. 1–12. Saṃ. 1915 = A.D. 1858 (with ṭīkā of Viśvanātha).
Vārāṇasī 35772 ff. 1–6. Saṃ. 1919 = A.D. 1862.
Goṇḍal 158 ff. 10–13. Saṃ. 1931 = A.D. 1874.
SOI 3466 4 ff. Saṃ. 1931 = A.D. 1874.
Vārāṇasī 35332 ff. 1–4. Saṃ. 1931 = A.D. 1874. Identical with Benares (1878) 155. 4 ff. Saṃ. 1931 = A.D. 1874. Apparently identical with Benares (1869) XXXVI 3. 4 ff. Saṃ. 1885 = A.D. 1828 (!).
Vārāṇasī 37174 ff. 1–10. Saṃ. 1933 = A.D. 1876.
Goṇḍal 159 ff. 11–36. Saṃ. 1945 = A.D. 1888.
Ānandāśrama 2047.
Ānandāśrama 2059.
Ānandāśrama 2144.
Ānandāśrama 2173.
Ānandāśrama 2544.
Ānandāśrama 2615.
Ānandāśrama 2617.
Ānandāśrama 2760.
Ānandāśrama 5402.
Ānandāśrama 5502.
Ānandāśrama 5635.
Ānandāśrama 5854.
Ānandāśrama 5857.
Ānandāśrama 5946.
Ānandāśrama 6510.
Ānandāśrama 6672.
Anup 4731 20 ff.
Anup 4732 12 ff.
ASBengal 6880 (G 10233) 3 ff. (*Pañcāṅgasādhana*)
ASBombay 238 3 ff.
ASBombay 239 4 ff.
Baroda 3282 4 ff.
Baroda 12622 6 ff.
BM 463 (Add. 14,365 e*) 8 ff.
Bombay U 351 7 ff.
Bombay U 352 2 ff.
Bombay U 353 8 ff.
Bombay U 354 4 ff.
BORI 902 of 1884/87 32 ff.
BORI 876 of 1887/91 11 ff. (with ṭīkā).
BORI 898 of 1891/95 3 ff.
BORI 177 of Viśrambag II 173 ff. (*sic!*).
Calcutta Sanskrit Coll. 121 4 ff.
CP, Hiralal 2014 (from Malkāpur in Bulḍānā District)
CP, Hiralal 2015 (from Jāvalabutā in Bulḍānā District)
CP, Hiralal 2016 (from Mulekheḍī in Bulḍānā District)
CP, Hiralal 2017 (from Bāsim in Akolā District)
CP, Hiralal 2018 (from Bāsim in Akolā District)
CP, Hiralal 2019 (from Mangrulpīr in Akolā District)
CP, Hiralal 2020 (from Ghuikheḍ in Amraotī District)
CP, Hiralal 2021 (from Brahmapurī in Chāndā District)
CP, Hiralal 2022 (from Nāgbhīḍ in Chāndā District)
CP, Hiralal 2023 (from Jubbulpore)
CP, Hiralal 2024 (from Jubbulpore)
CP, Hiralal 2025 (from Ratanpur in Bilāspur District)
CP, Hiralal 2026 (from Kārañjā in Akolā District)
CP, Hiralal 2028 (from Murtizāpur in Akolā District) (with ṭīkā)
CP, Hiralal 2029 (from Kholāpur in Amraotī District)
CP, Hiralal 2030 (from Mangrulpīr in Akolā District)
CP, Kielhorn XXIII 56 9 ff. (from Chāndā) (with ṭīkā of Viśvanātha).
DC 371 6 ff.
DC 385 3 ff.
DC 3297 4 ff.
DC 3323 15 ff.
DC 3884 13 ff.
DC 3885 ff. 5–35.
IO 2942 (2000) 28 ff.
IO 2943 (1990b) 3 ff.
Kotah 170 13 pp.
Nagpur 819 (1160) 5 ff.
Nagpur 821 (1706) 8 ff.
N-W Prov. I 86 35 ff. (from Benares) (*Pañcāṅgasādhana*).
PL, Buhler 183 10 ff. (with ṭīkā of Viśvanātha).
*Poleman 4801, 4800, 4802 (Smith Indic 129 A) ff. 1–33.
*Poleman 4854 (U Penn 709) 12 ff.
*Poleman 4855, 4861, 4892, 5125, 4951 (Smith Indic 138) ff. 1–23 and 27.
*Poleman 4856 (U Penn 1799) 27 ff.
*Poleman 4905 (U Penn 1847) 23 ff.
*Poleman 4907 (U Penn 1848) 23 ff.
*Poleman 4945 (Smith Indic 194) 18 ff.
RORI 4762 18 ff.

SOI 163.
SOI 2646 9 ff.
SOI 3336 12 ff.
SOI 5785.
SOI 7972.
SOI 9922.
SOI 10056 (with ṭīkā).
Tanjore D11441 8 ff.
Vārāṇasī 34356 ff. 1–11.
Vārāṇasī 34357 ff. 1–6.
Vārāṇasī 34651 ff. 1–5.
Vārāṇasī 35142 ff. 1–14 (with ṭīkā).
Vārāṇasī 35330 2 ff.
Vārāṇasī 35331 ff. 1–4. Probably identical with Benares (1878)77. 4 ff. Probably identical with Benares (1869) XV 5 4 ff.
Vārāṇasī 35334 ff. 1–13. Probably identical with Benares (1878) 79 15 ff. Probably identical with Benares (1869) XV 7 17 ff.
Vārāṇasī 35335 ff. 1–3.
Vārāṇasī 35752 ff. 1–4.
Vārāṇasī 36186 ff. 10–14 and 16–36.
Vārāṇasī 36665 ff. 1–2 and 4.
Vārāṇasī 36901 ff. 1–14.
VVRI 4743 6 ff. (with ṭīkā).

III. Tables of the *Tithicintāmaṇi:*

1. Table of tithis for 0 to 400 tithis, in 4 columns. Column 1 gives the tithi-number (koṣṭhaka); column 2 gives the day (modulo 7) and sexagesimal fraction of a day at which the tithi begins, assuming only the variation caused by solar motion (tithivārādi); column 3 lists the corrections in sexagesimal fractions of a day which must be applied to the entries in column 2 because of lunar motion (parākhya); and column 4 gives a number which, when divided into the so-called kendra, provides the number of ghaṭikās before or after sunrise that the tithi begins (hāra). The epoch values are: for tithivārādi 0;10,43 days; for parākhya, +0;0, 55 days; and for hāra +0;10 days. The epoch date is, in Śaka 1447 (A.D. 1525), Saturday 25 March; the conjunction of the Sun and Moon in fact occurred on 24 March 1525 at about Aries 8°.

Manuscripts: 4678 (Smith Indic 94 ff. 4v–16); 4801 (Smith Indic 129A ff. 1–11v); 4814 ff. 1v–5; 4854 ff. 1v–5v; 4855 (Smith Indic 138 ff. 1v–8v); 4856(a) ff. 1v–6; 4905 ff. 1v–10; 4906 ff. 1v–5v; 4907 ff. 1v–9v; 4945 ff. 1v–6v.

1a. Table of the parveśa and vya gu according to the Saurapkṣa, Keśavapakṣa, Āryapakṣa, and Brahmapakṣa. Not a table of the *Tithicintāmaṇi,* but corrections thereto. *Cf.* tables 8–10.

	Parveśa	Vya gu
Saurapakṣa	+0;0,40	+0;1,59
Keśavapakṣa	0;0,0	0;0,0
Āryapakṣa	+0;0,2	+0;0,5
Brahmapakṣa	−2;0,5	−0;0,16

The Saurapakṣa is the school of the *Sūryasiddhānta;* the Keśavapakṣa the school of the *Keśavīya* of Keśava, Gaṇeśa's father; the Āryapakṣa the school of the *Āryabhaṭīya* of Āryabhaṭa I; and the Brahmapakṣa the school of the *Brahmatulya* of Bhāskara II.

Manuscripts: 4801 (Smith Indic 129A f. 11v).

2. Table of nakṣatras for 0 to 390 nakṣatras, in 3 columns. Column 1 lists the nakṣatra-numbers (koṣṭhaka); column 2 gives the day (modulo 7) and sexagesimal fraction of a day at which the Moon enters the nakṣatra (nakṣatravārādi); and column 3 gives the divisor whose use yields the number of ghaṭikās before or after sunrise that the Moon enters the nakṣatra (hāra). The epoch positions are: for the nakṣatravārādi, 6;59,44 days; and for the hāra, +0;12,30 days.

Manuscripts: 4681 (Smith Indic 94 (b) ff. 1–12); 4800 (Smith Indic 129A ff. 12v–21v); 4814 ff. 5v–8; 4854 ff. 6–9; 4856 (b) ff. 1–6v; 4861 (Smith Indic 138 ff. 9v–14); 4905 ff. 10–15; 4906 ff. 6–10; 4907 ff. 10–15; 4945 ff. 7–12.

3. Table of yogas for 0 to 420 yogas in 4 columns. Column 1 lists the yoga-numbers (koṣṭhaka); column 2 gives the day (modulo 7) and sexagesimal fraction of a day on which the yoga begins, assuming only the variation caused by solar motion (yogavārādi); column 3 gives the corrections to be applied to the entries in column 2 because of lunar motion (parākhya); and column 4 again lists the divisors by which one finds out the number of ghaṭikās before or after sunrise that a yoga begins (hāra). The epoch positions are: for the yogavārādi, 6;51,1 days; for the parākhya, −0;0,46 days; and for the hāra, +0;12 hours.

Manuscripts: 4682 (Smith Indic 94 (c) ff. 1–12); 4802 (Smith Indic 129A ff. 22v–32v); 4814 ff. 8–12; 4854 ff. 9v–12v; 4856 (c) ff. 1–6; 4892 (Smith Indic 138 ff. 15–23v); 4905 ff. 15–23; 4906 ff. 10v–15; 4907 ff. 15v–23v; 4945 ff. 13–18v.

4. Table of days and day-fractions on which the Sun enters each of the 27 nakṣatras. *Cf.* table 16 of the *Jagadbhūṣaṇa.*

Aśvinī	4;49,34a	Svāti	2;26,56
Bharaṇī	4;29,41	Viśākhā	1;44,44
Kṛttikā	4;17,27	Anurādhā	0;56,56
Rohiṇī	4;10,43	Jyeṣṭhā	0;4,31
Mṛgaśiras	4;7,23	Mūla	6;8,46
Ārdrā	4;11,43	Pūrvāṣāḍhā	5;10,44
Punarvasu	4;16,53	Uttarāṣāḍhā	4;12,13
Puṣya	4;17,40	Śravaṇa	3;14,58
Āśleṣā	4;20,48	Dhaniṣṭhā	2;20,19
Maghā	4;17,37	Śatabhiṣak	1;29,38
Pūrvaphālgunī	4;8,57	Pūrvabhādrapadā	0;43,56
Uttaraphālgunī	3;53,54	Uttarabhādrapadā	0;4,12

Hasta	3;31,51	Revatī	6;30,49
Citra	3;2,48		

a var. 6;4,34

Manuscripts: 4814 f. 12v; 4856 (d) f. 1v; 4906 f. 15; 5125 (Smith Indic 138 f. 27).

5. Table of days and day-fractions on which the Sun enters each of the 12 zodiacal signs. *Cf.* table 16 of the *Jagadbhūṣaṇa.*

Aries	4;49,34a	Libra	2;45,47
Taurus	0;44,57	Scorpio	4;39,21
Gemini	4;10,50	Sagittarius	6;8,46
Cancer	0;48,47	Capricorn	0;27,23
Leo	4;17,37	Aquarius	1;54,28
Virgo	0;19,4	Pisces	3;43,33

a. var. 6;4,34

Manuscripts: 4814 f. 12v; 4856 (d) f. 1v; 4906 f. 15; 4945 f. 18v; 4951 (Smith Indic 138 f. 27v).

The relationship of the next 2 tables (6-7) to the *Tithicintāmaṇi* is not certain. They are probably not part of the original, though related to it.

6. Table involving the 27 nakṣatras and corresponding ghaṭikās. Its purpose is not understood.

Manuscripts: 4856 (c) f. 7; 4906 ff. 15–15v.

7. Corrections to tables 1 to 3 for differences in time and latitude. There are 7 columns. Column 1 gives the argument, in steps of 10, from 0 to 400. Column 2 gives the tithi ayana, column 3 its palabhāntara; columns 4 and 5 similarly are devoted to the nakṣatra ayana and its palabhāntara, and columns 6 and 7 to the yoga ayana and its palabhāntara. Again the meaning remains obscure.

Manuscripts: 4856 (c) ff. 7–7v.

Tables 8, 9, and 10 are not from the *Tithicintāmaṇi.*

8. Table of parameters for (1) the Lord of the Year (abda), (2) the tithidhruva, (3) the nakṣatrayogadhruva, (4) the tithikendra, (5) the nakṣatrakendra, (6) the yogakendra, (7) the tithibhoga, (8) the nakṣatrabhoga, (9) the yogabhoga, and (10) the vya gu (?).

	1	2	3	4
cycles	7	30	27	27;59,33
yearly cālaka	0;15,31,31	11	10	7;5,56
āryapakṣe tithicālaka	−0,6,6	0,0,0	0,0,0	+0,7,26
brahmapakṣe	−0,3,7	0,0,0	0,0,0	+0,23,25
Keśavīye	+23	0,0	0,0	+0,17,59

	5	6	7	8	9	10
cycles	27;13,48	29;16,1	7	7	7	12
cālaka	7;0,56	7;31,57	1;11,41	1;18,3	1;7,52	
āryapakṣe	+0,8,40	+0,6,40	+0,1,39	+0,1,4	+0,0,34	
brahmapakṣe	+0,26,46	+0,24,25	−0,2,11	−0,2,15	−0,2,19	
Keśavīye	+0,17,22	+0,18,43	−0,0,15	−0,0,12	−0,0,10	

The parameters for cycles and cālakas are those of the Anonymous of 1741.

Manuscripts: 4856 (d) f. 1. *Cf.* 4945 f. 12v.

9. Table of the motions of the same 10 functions whose parameters were given in table 8, plus an 11th: the parveśa. These tables are for 1 to 10 periods of 46 years.

Manuscripts: 4856 (d) ff. 2–2v.

10. Tables of the motions of the same 11 functions which were tabulated in table 9. These tables are for 0 to 46 years. The epoch values are:

Lord of the Year	3;44,38
Tithidhruva	4
Nakṣatrayogadhruva	5
Tithikendra	2;19,46
Nakṣatrakendra	2;44,24
Yogakendra	2;53,8
Tithibhoga	4;4,53
Nakṣatrabhoga	4;33,43
Yogabhoga	4;30,19
Vya gu	7;22,23
Parveśa	6;17,28

Manuscripts: 4856 (d) ff. 3–8.

THE *GRAHALĀGHAVĪYA-MADHYAMASPAṢṬĀRKASĀRIṆĪ*

I. Manuscripts:

*Poleman 4819 (Harvard 59) 7 ff. Saṃ. 1866 = A.D. 1809.

II. Tables:

This work is based on Gaṇeśa's *Grahalāghava* (1520), a very popular textbook which was frequently commented on and has often been published. There are many sets of tables based on its parameters—e.g., by Premamiśra (1656); *cf.* the *Grahalāghavasāriṇī* analyzed below. The epoch of the present work is virtually identical with that of table 5 of the *Tithicintāmaṇi*: Wednesday 22 March 1525.

The only table is of the mean longitudes, true longitudes, and daily progresses of the Sun for 0 to 365 days. The mean longitude is 0;0,0,0° on day 0; 5,59;44, 42° on day 365. The maximum equation is 2;10,44°. The maximum daily progress is 1;1,23° on days 260–263; the minimum 0;56,53° on days 79–80. The unique manuscript, on f. 1r, reads: śākaḥ 1447 madhyamameṣārkasamaye 'bdapaḥ 4;45,27 prativarṣaṃ guṇakāḥ 1;15,31,32,46.

Manuscripts: 4819 ff. 1–7v.

THE *BṚHATTITHICINTĀMAṆI* OF GAṆEŚA

I. Manuscripts:

Vārāṇasī 35725 ff. 1–7. Saṃ. 1722 = A.D. 1665.

Baroda 3389 19 ff. Saṃ. 1811 = A.D. 1754. (with ṭīkā of Viṣṇu).

*Poleman 4709 (Smith Indic 151) 107 ff. Śaka 1682 = A.D. 1761.

BORI 428 of 1895/98 31 ff. Saṃ. 1839 = A.D. 1782.

*Poleman 4680 (Smith Indic 92) ff. 5–12. Śaka 1704 = A.D. 1782.
Baroda 3205 6 ff. Saṃ. 1886 = A.D. 1829.
Baroda 9454 23 ff. Śaka 1783 = A.D. 1861 (with ṭīkā of Viṣṇu).
Alwar 1871 (with ṭīkā of Viṣṇu).
ASBombay 237 8 ff.
BM 460 (Add. 14,365m) 6 ff.
BM 461 (Add. 14,363 f) 1 f.
Kotah 159 13 pp.
*Poleman 4708 (Smith Indic 16) 39 ff.
RORI 5747 21 ff. (with ṭīkā of Viṣṇu).

II. Tables:

The *Bṛhattithicintāmaṇi* was a less popular work of Gaṇeśa; manuscripts of it are generally confined to Western India, and it was commented on only once, by Viṣṇu (fl. 1608). A truncated form of the text (without the tables) and the commentary of Viṣṇu have been published. The epoch of the text is Śaka 1474 (A.D. 1552).

The idea of the work is to tabulate the true tithivāras, nakṣatravāras, and yogavāras for 27 cycles each. Each cycle consists of 0 to 371 tithis, 0 to 362 nakṣatras, and 0 to 389 yogas—i.e., each is approximately 1 year. This arrangement permits one to avoid the difficulties of the *Tithicintāmaṇi*-tables, in which the two variants due to lunar and solar motion are kept separate, by substituting a single function which is approximately cyclical.

1. Table of 27 cycles of the tithikendra.
Manuscripts: 4709 (b) ff. 1–40; *cf.* 4708 ff. 1–39v.

2. Table of 27 cycles of the nakṣatrakendra.
Manuscripts: 4709 (c) ff. 1–32v; *cf.* 4708 ff. 1–39v.

3. Table of 27 cycles of the yogakendra.
Manuscripts: 4709 (d) ff. 1–33; *cf.* 4708 ff. 1–39v.

THE *TITHYĀDICINTĀMAṆI* OF DINAKARA

I. The Life of Dinakara:

The name of the author is given in the first verse of the *Tithyādicintāmaṇi:*

śrīsūryapramukhān grahān vidhiharīśān vighnarājyaṃ giraṃ
bhaktyā namya guroḥ padābjayugalaṃ siddhāntavid vāḍabān /
dṛṣṭvā vai racitaṃ sphuṭaṃ ca sugamaṃ yāmārdhasādhyaṃ tithis-
pattraṃ yena karomy ahaṃ dinakaras tithyādicintāmaṇim // 1 //

Further details are given in verses 11 and 12:

śrīmaty unnatadurgānāṁni nagare jyotirvidāṃ bhāskaro
vāyusthāpitavipravaṃśatilakaḥ śrīśoṣaṇākhyo dvijaḥ /
śrautasmārtavicārasāracaturaḥ śrīśaṅkaropāsakaḥ
kāśīdvāravatīgayātripathigātīrthāśrayaḥ satyavāk
// 11 //
putras tasya tadaṅghripadmayugajaprāptaprasādaḥ sudhīr
varṣe rāmayugāṅgabhūparimite śrīvikramārkād gate /
śrutyādyācyutavāsare (?) dinakaraḥ śrīrāmacandrāgajo (?)
vijñas tena kṛto budhaiḥ karuṇayā tithyādicintāmaṇiḥ
// 12 //

From this it follows that the *Tithyādicintāmaṇi* was composed in Saṃ. 1626 (A.D. 1569) by Dinakara, the son of the pious Śoṣaṇa of Unnatadurgā (Uparkot in Junāgaḍh, Saurāṣṭra?).

II. Manuscripts:

Baroda 3154 3 ff.
*Poleman 4718 (Smith Indic 53) 2 ff.

III. Tables:

1. Table of tithivāras for 1 to 13 months.
Manuscripts: 4718 f. 1v.

2. Table of yogavāras for 1 to 15 months.
Manuscripts: 4718 f. 2.

3. Table of nakṣatravāras for 1 to 14 months.
Manuscripts: 4718 f. 2.

THE *CANDRĀRKĪ* OF DINAKARA

I. Life of Dinakara:

Dinakara wrote two works on astronomy: the *Kheṭakasiddhi* on the five star-planets, and the *Candrārkī* on astronomical matters connected with the two luminaries. From manuscripts of the first work (BORI 303 of 1882/83, on which see R. G. Bhandarkar, *Report of the Search for Sanskrit MSS. in the Bombay Presidency during the Year 1882–83,* Bombay 1884, pp. 27–28; and IO 2947 [2648]) one learns that he was the great-grandson of Dunda Moḍha of the Kauśikagotra from Bārejya or Bāreja, perhaps Bariya in Rewa Kantha, Gujarat. A manuscript of the *Candrārkī* (Baroda 3119) informs us that his father's name was Rāmeśvara. The last verse of the *Candrārkī* in one of our manuscripts (4716) reads as follows:

bārejākhyavasagrāme cakre dinakaro muhā (?) /
jātaḥ kauśikagotre ca mauḍhajñātisamudbhavaḥ//38//

There is also a report, recorded by Bhandarkar, that he lived in a village named Jñāvāra, which has been tentatively identified with Budwarpet in Poona. As the epoch of both of his works is Śaka 1500, Saṃvat 1635 (A.D. 1578), he evidently flourished in the second half of the sixteenth century.

II. Manuscripts:

A. *Kheṭakasiddhi:*

Goṇḍal 35 8 ff. Saṃ. 1793 = A.D. 1736.
BORI 303 of 1882/83 6 ff. Saṃ. 1796 = A.D. 1739.
Gujarāt Vidyā Sabhā 3157 ff. 1–18. Saṃ. 1799 = A.D. 1742. (*Dinakarasāraṇī*)

RORI 619 31 ff. Saṃ. 1875 = A.D. 1818 (*Dinakara-sāraṇī*)
Anup 4503. 83 ff. Probably identical with PL, Buhler 45 84 ff.
Baroda 1081 5 ff.
IO 2947 (2648) 91 ff.
Jaipur 3 ff.
Jaipur 18 ff. (*Dinakarīsāraṇī*)
RORI 4731 30 ff.

B. *Candrārkī.*

Goṇḍal 77 28 ff. Saṃ. 1737, Śaka 1602 = A.D. 1680.
PL, Buhler 90 4 ff. Saṃ. 1738 = A.D. 1681.
Goṇḍal 80 6 ff. Saṃ. 1745 = A.D. 1688 (ravipañcāṅga only)
LDI 3106 12 ff. Saṃ. 1781 = A.D. 1724.
Goṇḍal 82 4 ff. Saṃ. 1814 = A.D. 1757.
Goṇḍal 78 17 ff. Saṃ. 1820, Śaka 1686 = A.D. 1764.
RORI 2582 6 ff. Saṃ. 1828 = A.D. 1771 (with ṭīkā).
*Poleman 4827 (Smith Indic 180) 2 ff. Saṃ. 1829, Śaka 1694 = A.D. 1772.
RORI 4870 3 ff. Saṃ. 1839 = A.D. 1782.
LDI 4331 15 ff. Saṃ. 1844 = A.D. 1787.
LDI 7834 20 ff. Saṃ. 1844 = A.D. 1787.
Goṇḍal 79 12 ff. Saṃ 1853 = A.D. 1796. (with ṭīkā).
Goṇḍal 84 3 ff. Saṃ. 1857 = A.D. 1800.
Goṇḍal 85 2 ff. Saṃ. 1878, Śaka 1743 = A.D. 1821.
Gujarāt Vidyā Sabhā 4198 6 ff. Saṃ. 1885 = A.D. 1828.
Anup 4566 1 f. Saṃ. 1903 = A.D. 1846.
BORI 510 of 1895/1902 23 ff. Saṃ. 1904 = A.D. 1847 (with *Jātakapaddhati*).
RORI 2584 4 ff. Saṃ. 1904 = A.D. 1847.
Goṇḍal 83 3 ff. Saṃ. 1916 = A.D. 1859.
Goṇḍal 81 4 ff. Sam. 1937 = A.D. 1880.
Goṇḍal 86 2 ff. Saṃ. 1970 = A.D. 1913.
Adyar Index 2019.
Baroda 3119 7 ff. (with ṭīkā).
Baroda 3120 3 ff.
BORI 455 of A 1881/82 7 ff. (māsapraveśasāraṇī only)
BORI 308 of 1882/83 4 ff.
BORI 315 of Viśrambag I 4 ff.
GOMLMadras 14033 38 pp.
Gujarāt Vidyā Sabhā 4203 ff. 1–4.
Gujarāt Vidyā Sabhā 4491 ff. 1–3.
Gujarāt Vidyā Sabhā 4577 ff. 1, 3–4, and 2–3.
Gujarāt Vidyā Sabhā 5258 ff. 1–4.
IO 2948 (2541e) 2 ff. (ṭīkā).
Jaipur 10 ff.
Kotah 161 3 pp.
LDI 4159 13 ff.
LDI 4163 8 ff.
LDI 6570 10 ff.
LDI 6931 11 ff.
LDI 7031 7 ff.
LDI 7226/1 ff. 1v–2 (with ṭīkā)
LDI 7401/2 ff. 5–6v.
Oxford 775 (Walker 208b) ff. 59-65.
Paris BN 1005 (Sans. Dév. 331–340) VIII.
PL, Buhler 91 49 ff. (ṭīkā).
PL, Buhler 92 22 ff. (ṭīkā).
*Poleman 4716 (Harvard 525) 5 ff.
*Poleman 4717, 4923, 4824, 4823 (Smith Indic 190) ff. 3–6 and 8–17.
*Poleman 4825 (Smith Indic 58) 15 ff.
*Poleman 4826 (Harvard 934) 14 ff.
*Poleman 4883 (Smith Indic 34) ff. 9–11.
*Poleman 4895 (Smith Indic 40) 4 ff.
*Poleman 4946 (Smith Indic MB) XXIV f. 19.
*Poleman 4946 (Smith Indic MB) XXXVIII 1 f.
*Poleman 4946 (Smith Indic MB) XXXIX 1 f.
*Poleman 4946 (Smith Indic MB) LII 2 ff.
*Poleman 4946 (Smith Indic MB) LXXXVIII 1 f.
*Poleman 4946 (Smith Indic MB) LXXXIX f. 1.
*Poleman 4946 (Smith Indic MB) XCI ff. 1–2.
*Poleman 4946 (Smith Indic MB) XCII ff. 1–2.
*Poleman 4949 (Smith Indic 19) 2 ff.
*Poleman 4952 (Smith Indic 29) 6 ff.
*Poleman 5178 (Smith Indic 35) 11 ff.
*Poleman 5179 (Smith Indic 46) ff. 2–10.
RORI 224 2 ff.
RORI 3253 1 f.
RORI 3815 2 ff.
RORI 4745 14 ff.
RORI 4893 3 ff.
SOI 9467.
Vārāṇasī 35035 ff. 1–4 and 4–9.

III. Tables:

1. Table of longitudes of the Sun and its daily progresses for 0 to 365 days. The initial longitude, when the mean Sun is at Aries 0°, is 2;9,13°. The maximum daily progress is 1;1,23° at 259 to 266 days; the minimum 0;56,54° at 73 to 86 days.

Manuscripts: 4825 ff. 1–5v; 4826 ff. 1v–5; 4895 ff. 1–4v; 4923 (Smith Indic 190 ff. 3v–6v); 4946 LXXXIX ff. 1–1v; 5178 ff. 1v–4v; 5179 ff. 2–4.

2. Table of mean longitudes of the Moon and its Anomaly for 0 to 365 days. The initial longitudes are respectively 5,59;54,17° and 5,59;54,20°; *cf. Jagadbhūṣaṇa* tables 10 and 11. The daily mean motions of the Moon and its Anomaly are respectively 13;10,34,53° and 13;3,54,10°.

Manuscripts: 4824 (Smith Indic 190 ff. 8–12v); 4825 ff. 6–12v; 4826 ff. 5v–9; 4946 LXXXVIII f. 1; 4952 ff. 1–2v; 5178 ff. 4v–7v; 5179 ff. 4v–7v.

3. Table of mean motions of the Moon and its Anomaly for 1 to 60 ghaṭikās. The mean daily motions are respectively 13;10,35° and 13;3,54°. *Cf.* table 4 of the Anonymous of 1704, and table 14 of the *Jagadbhūṣaṇa*.

Manuscripts: 4826 f. 9v; 4946 XXIV ff. 19–19v; 4952 ff. 2v–3; 5178 f. 8; 5179 f. 8.

6. Double-entry table of māsapraveśa and dinapraveśa for 0° to 29° horizontal, and for Aries to Pisces verti-

cal. For each degree of solar motion two things are noted: the days (modulo 7) to the beginning of the next solar month—i.e., the days it will take the Sun to travel 30° from that degree for which the table is entered; and the ghaṭikās that must be added algebraically to 1 day to give the time it will take the Sun to travel 1°. For the first entry, the māsapraveśa, the maximum is 3;37,6 days (i.e., 31;37,6 days) at Gemini 1° to 2°, and the minimum 1;19,44 days (i.e., 29;19,44 days) at Sagittarius 1° to 2°.

Manuscripts: 4946 XCI ff. 2–2v; 4946 XCII ff. 1–2v; 4949 ff. 1–1v.

7. Table of week-days corresponding to the beginning of the first tithi, nakṣatra, and yoga of the years 1 to 84 from the epoch (i.e., for A.D. 1579 to 1662). The epoch positions are:

for tithis		for nakṣatras		for yogas	
gaṇa	47	gaṇa	64	gaṇa	74
tithi	4	nakṣatra	3	yoga	3
day	1;18	day	1;13	day	1;42

The significance of the gaṇa is not clear to me.

Manuscripts: 4946 XCI ff. 1–1v; 4949 f. 2.

4. Table of the equation of the Moon for 0° to 359° of anomaly; also given are the differences between successive equations, and the daily progresses of the Moon. The maximum equation is 5;2,35° at 90° and 270°. The maximum daily progress is 14;19,24° at 180°; the minimum 12;1,46° at 0°.

Manuscripts: 4823 (Smith Indic 190 ff. 13–17v); 4825 ff. 13–15v; 4826 ff. 10–14; 4883 ff. 9–11; 4952 ff. 3–6; 5178 ff.8v–11v; 5179 ff. 8v–10.

5. Tables of the mean motions of the Lord of the Year, Epact, Moon, and Lunar Anomaly for 1 to 20 years and for 20 to 400 years in steps of 20. The epoch positions are added to the first entries of the collected-year (labdha) tables; they indicate that the epoch is Śaka 1500 (A.D. 1578). All parameters are derived from the *Mahādevī* where possible.

	Yearly motions	Epoch positions
Lord of the Year	1;15,31,17,17 days	1;24,57,12,30 days
Epact	11;3,53,22,40 tithis	22;18,19,10,0 tithis
Moon	2,12;46,40,32°	4,27;39,50,0°
Anomaly	1,32;6,8,48°	4,21;42,23,8°

Manuscripts: 4827 ff. 1–2v. *Cf.* 4946 XXXIX f. 1 and 4946 LII ff. 1–2v.

ANONYMOUS OF 1578

I. Manuscripts:

*Poleman 4863 (Smith Indic 81) 3 ff.

II. Tables:

1. Tables of mean motions for the Lunar Node, Mars, Mercury, Jupiter, Venus, and Saturn, at intervals twice as large as those of the *Mahādevī,* but using *Mahādevī* parameters; in both labdha and śeṣa tables, k = 1 to 30. Thus the set of tables of true longitudes consists of all those tables in which N equals 0 or an even number. The epoch of these tables is Śaka 1500 (A.D. 1578).

	epoch position	epoch × 2	computed from *Mahādevī* (for Śaka 1500)
Lunar Node	1;17,46,27,49 (× 12°)	2;35,32,55,38 (× 6°)	– 5,44;26,42,26,12°
Mars	21;5,39,52,45	42;11,19,45,30	42;11,19,45,30 (× 6°)
Mercury	20;45,22,40,23	41;30,45,20,46	41;30,45,20,46
Jupiter	14;22,2,14,15	28;44,4,28,30	28;44,4,28,30
Venus	28;4,47,56,30	56;9,35,53,0	56;9,35,53,0
Saturn	22;34,2,13,0	45;9,4,26,0	45;9,4,26,0

ANONYMOUS II OF 1578

I. Manuscripts:

*Poleman 4917 (Smith Indic 25) f. 1.

II. Tables:

1. Table of the Epact for Śaka 1501 to 1660 (A.D. 1579 to 1738). The yearly parameter is about 11;3,53,30 tithis. The epoch position for Śaka 1501 (A.D. 1579) is 3;22,13 tithis; the position computed from the *Mahādevī* is 3;22,12,32,40 tithis.

Manuscripts: 4917 f. 1. *Cf.* 4946 LI f. 1.

2. Table of intercalary months from Śaka 1500 to 1666 (A.D. 1578 to 1744) according to a 19-year cycle.

1500	Āśvina	1560	Bhādrapada (?)
1503	Bhādrapada	1563	Caitra
1506	Āṣāḍha	1566	Vaiśākha
1509	Vaiśākha	1568	Śrāvaṇa
1511	Bhādrapada	1571	Āṣāḍha
1514	Śrāvaṇa	1574	Vaiśākha
1517	Jyeṣṭha	1576	Bhādrapada
1519	Āśvina	1579	Śrāvaṇa
1522	Bhādrapada	1582	Caitra
1525	Āṣāḍha	1585	Vaiśākha
1528	Caitra	1587	Śrāvaṇa
1530	Bhādrapada	1590	Śrāvaṇa (?)
1533	Śrāvaṇa	1593	Vaiśākha
1536	Caitra	1595	Bhādrapada
1538	Āśvina	1598	Bhādrapada (?)
1541	Śrāvaṇa	1601	Jyeṣṭha
1544	Āṣāḍha	1603	Āśvina
1547	Caitra	1606	Śrāvaṇa
1549	Śrāvaṇa	1609	Āṣāḍha
1552	Āṣāḍha	1612	Vaiśākha
1555	Vaiśākha	1614	Bhādrapada (1615 in *ms*)
1557	Bhādrapada	1617	Āṣāḍha

1620	Jyeṣṭha	1644	Śrāvaṇa
1622	?	1647	Āṣāḍha
1625	Śrāvaṇa	1650	Vaiśākha
1628	Āṣāḍha	1652	Bhādrapada
1631	Caitra	1655	Śrāvaṇa
1633	Bhādrapada	1658	Jyeṣṭha
1636	Āṣāḍha	1660	Āśvina
1639	Vaiśākha	1663	Āṣāḍha
1642	Bhādrapada	1666	Vaiśākha

Manuscripts: 4917 f. 1v.

ANONYMOUS III OF 1578

I. Manuscripts:

*Poleman 4763a (Smith Indic 41) Mercury f. 31.
*Poleman 4868 (Smith Indic 140) D f. 1.

II. Tables:

1. Table of the motion of the Lunar Node for periods of 20 years from Śaka 1500 to 1700 (A.D. 1578 to 1778), and for 1 to 20 years. The yearly mean motion is -19;21,37°; the epoch position for Śaka 1500 (A.D. 1578): 5,44;26,43°. *Cf.* table 5 of the *Candrārkī.*

Manuscripts: 4868 D f. 1.

2. Table of the motion of the Lunar Node for 1 to 27 avadhis. *Cf.* table 15 of the *Jagadbhūṣaṇa.*

Manuscripts: 4763a Mercury f. 31; 4868 D f. 1.

ANONYMOUS OF 1594

I. Manuscripts:

*Poleman 4942 (Smith Indic 119) ff. 2–18.
*Poleman 4945 (Smith Indic 194) f. 12.
*Poleman 4946 (Smith Indic MB) XLVI f. 6.

II. Tables:

1. Table of parameters of yearly motions:

Epact	11;3,53,22,39 tithis
Lord of the Year	1;15,31,17,17 days
Tithikendra	7;9,41,52,24 tithis modulo 27;59,33
Nakṣatrakendra	6;58,0,14,23 days modulo 27;13,48
Yogakendra	7;29,15,56,13 days modulo 29;16,1

Manuscripts: 4942 f. 2.

2. Table of epoch positions:

Epact	20;21,48,29,5
Lord of Year	0;33,19,32,26
Tithikendra	23;1,39,16,42
Nakṣatrakendra	22;26,35,42,52
Yogakendra	24;5,36,27,38

The positions of the epact and the lord of the year indicate a date 278 years after the epoch of the *Mahādevī* —i.e., Śaka 1516 (A.D. 1594).

Manuscripts: 4942 f. 2.

3. Table of week-days on which the Sun enters each of the zodiacal signs, of the Sun's daily progress at that time, of the length of daylight on that day, and of a cālaka.

Aries	4;46,58	0;58,27°	31;12	+3½
Taurus	0;44,23	0;57,30	32;56	+2
Gemini	4;9,32	0;57,0	34;0	-1
Cancer	0;46,46	0;57,3	33;52	-2½
Leo	4;15,3	0;57,53	33;38	-3½
Virgo	0;16,58	0;59,8	30;48	-4
Libra	2;43,50	1;0,0	28;48	-3½
Scorpio	4;37,0	1;1,7	27;4	-2
Sagittarius	6;5,53	1;1,22	26;0	-1
Capricorn	0;25,25	1;1,7	26;8	+2½
Aquarius	1;52,4	1;0,27	27;22	+3½
Pisces	3;40,49	0;59,7	29;12	+4

Manuscripts: 4942 f. 2.

4. Table of week-days on which the Sun enters each of the 27 nakṣatras, of the Sun's progress on that day, and of the muhūrtas of the Sun's entry.

Aśvinī	4;49,28	0;58,20°	30
Bharaṇī	4;30,37	0;57,59	15
Kṛttikā	4;18,26	0;57,34	30
Rohiṇī	4;12,20	0;57,14	45
Mṛgaśiras	4;11,0	0;57,1	30
Ārdrā	4;13,2	0;56,55	15
Punarvasu	4;19,26	0;56,57	45
Puṣya	4;19,36	0;57,6	30
Āśleṣā	4;20,21	0;57,22	15
Maghā	4;17,12	0;57,44	30
Pūrvaphālgunī	4;8,47	0;58,11	30
Uttaraphālgunī	3;53,54	0;58,40	45
Hasta	3;32,13	0;59,10	30
Citra	3;3,17	0;59,42	30
Svāti	2;27,20	1;0,10	15
Viśākhā	1;44,56	1;0,37	45
Anurādhā	0;56,52	1;0,58	30
Jyeṣṭhā	0;4,15	1;1,13	15
Mūla	6;8,21	1;1,22	30
Pūrvāṣāḍhā	5;10,34	1;1,23	30
Uttarāṣāḍhā	4;12,22	1;1,20	45
Śravaṇa	3;15,17	1;1,8	30
Dhaniṣṭhā	2;10,37	1;0,48	30
Śatabhiṣak	1;29,48	1;0,23	15
Pūrvabhādrapadā	0;44,21	0;59,59	30
Uttarabhādrapadā	0;4,11	0;59,30	45
Revatī	6;31,24	0;58,58	30

Manuscripts: 4942 f. 2v.

5. Table of "kings" (rāja) for each zodiacal sign. Each of the 7 planets is rāja in each of the signs for a number (indicating time? or degrees?). The numbers employed are:

2,8; 2,11; 2,14; 5,2; 5,5; 5,14; 8,2; 8,8; 8,11; 8,14; 11,5; 11,8; 11,11; 11,14; 14,2; 14,11; and 14,14. The purpose of the table is obscure. *Cf.* Table 8 of the *Pañcāṅgavidyādharī,* and the *Bhāsvatī* of Śatānanda, edited by Mātṛprasāda Pāṇḍeya, Benares 1917, pp. 38–40.

Manuscripts: 4942 f. 2v; 4945 f. 12; 4946 XLVI f. 6.

6. Table for correcting the length of the tithikendra for 0 to 27 days of an anomalistic month horizontal, and for 0 to 59 ghaṭikās vertical; normed so as to be always positive. The maximum entry is 1;24,40, the minimum 0;35,20, and the mean 1;0,0.

Manuscripts: 4942 ff. 3–7v.

7. Table for correcting the length of the nakṣatra for 0 to 27 days of an anomalistic month horizontal, and for 0 to 59 ghaṭikās vertical; normed so as to be always positive. The maximum entry is 1;22,51, the minimum 0;37,9, and the mean 1;0,0.

Manuscripts: 4942 ff. 8–12v.

8. Table for correcting the length of the yoga for 0 to 29 days of a synodic month horizontal, and for 0 to 59 ghaṭikās vertical; normed so as to be always positive. The maximum entry is 1;21,17, the minimum 0;38,43, and the mean 1;0,0.

Manuscripts: 4942 ff. 13–17v.

ANONYMOUS OF 1598

I. Manuscripts:

*Poleman 4807 (Smith Indic 27) 1 f.

*Poleman 4946 (Smith Indic MB) XVI 1 f.

II. Tables:

1. Tables of mean motions of the Lord of the Year, Epact, Moon, Lunar Anomaly, Mars, Mercury, Jupiter, Venus, Saturn, and the Lunar Node for 1 to 20 years. *Cf.* table 5 of the *Candrārkī* and table 1 of Anonymous III of 1578.

Manuscripts: 4807 f. 1; 4946 XVI ff. 1–1v.

2. Tables of mean motions of the same entities for periods of 20 years from Śaka 1520 to 1700 (A.D. 1598 to 1778). All the parameters except for those for the Lunar Anomaly and the Lunar Node are taken from the *Mahādevī;* the parameter for the Lunar Anomaly is from the *Candrārkī*. It is likely that these tables, then, are based on those of the *Candrārkī* and *Kheṭakasiddhi*.

The epoch positions indicate a date of Thursday, 30 March 1598.

	Sidereal	Tropical	Sidereal-Tropical
Lord of the Year	5;35,22,58,10 days		
Epact	3;36,6,43,20 tithis		
Moon	43;13,20,40°	65°	−22°
Lunar Anomaly	5,3;45,19,0		
Mars	2,1;11,5,15,0	103	+18!
Mercury's anomaly	4,24;10,52,48,36	(Mercury) 352	
Jupiter	59;26,37,5,0	73	−14
Venus' anomaly	2,40;55,31,30,0	(Venus) 57	
Saturn	2,35;11,35,0,0	179	−24
Lunar Node	5,17;16,22,28,12	c.331	

Manuscripts: 4807 f. 1v.

THE *JAGADBHŪṢAṆA* OF HARIDATTA

I. The Life of Haridatta:

Haridatta, the son of Harajī, composed the *Jagadbhūṣaṇa* during the reign of Jagatsiṃha I (1628–1652) of Mewar. The epoch he used is Śaka 1560 (A.D. 1638).

II. Manuscripts:

LDI 6182 5 ff. Saṃ. 1717 = A.D. 1660.
Anup 4589 69 ff. Saṃ. 1720 = A.D. 1663.
RORI 4882 53 ff. Saṃ. 1737 = A.D. 1680.
LDI 5545 93 ff. Saṃ. 1759 = A.D. 1702.
*Poleman 4944 (Smith Indic 172) f. 151. Śaka 1626 = A.D. 1704.
RORI 3744 88 ff. Saṃ. 1788 = A.D. 1731.
Tod 36(5) 87 ff. Saṃ. 1820 = A.D. 1763.
RORI 4710 65 ff. Saṃ. 1829 = A.D. 1772.
*Poleman 4946 (Smith Indic MB) XXXIV f. 138. Saṃ. 1899, Śaka 1764 = A.D. 1842.
Anup 4586 12 ff.
Anup 4587 96 ff.
Anup 4588 73 ff.
BORI 823 of 1887/91 14 ff.
Jaipur 72 ff.
Jaipur 88 ff.
Jaipur 8 ff.
Jaipur 7 ff.
Kotah 166 95 pp.
Mitra, Not. 3118 8 ff.
PL, Buhler 99.
PL, Buhler 100.
*Poleman 4869 (Smith Indic 146) 100 ff.
RORI 3760 71 ff.
RORI 4727 60 ff.
RORI 4787 55 ff.
RORI 4894 4 ff.

III. Tables:

1–5. Planets. Tables of the true longitudes of the planets. As these are all computed according to the same general scheme, they will be discussed together here.

For each planet the tables are arranged in 3 columns. Column 1 contains the numbers of the avadhi (1 to 27). Column 2 lists the true longitudes. Column 3 gives the planet's daily motion at that longitude. Thus far, the set-up is comparable to that of the solar table (12). In the margins are added indications of the occurrences of the Greek-letter phenomena.

A cycle is employed for each planet similar to those

of the Babylonian goal-year texts (A. Sachs, "A Classification of Babylonian Astronomical Tablets of the Seleucid Period," *Jour. Cuneiform Studies* 2, 1950, 271–290), Ptolemy (*Syntaxis* 9,3), the *Vākyakaraṇa* (ed. T. S. Kuppanna Sastri and K. V. Sarma, Madras 1962), and al-Zarqālla (ed. J. M. Millás Vallicrosa, *Estudios sobre Azarquiel*, Madrid 1942–50, pp. 177–214; see the forthcoming paper by M. Boutelle, "The Almanac of Azarquiel").

	Babylonian	Ptolemy
Saturn	2 rev. in 59y.	2 rev. + 1;40° in 59y + 1;35d.
Jupiter	6 rev. in 71y.	6 rev. + 4;50° in 71y + 4;54d.
Mars	42 rev. in 79y.	42 rev. + 3;10° in 79y + 3;13d.
Venus	8 rev. in 8y.	8 rev. − 2;15° in 8y − 2;18d.
	(5 synodic periods)	(5 synodic periods).
Mercury	46 rev. in 46y.	46 rev. + 1° in 46y + 1;2d.
	(145 synodic periods)	(145 synodic periods).

	Vākyakaraṇa	al-Zarqālla
Saturn	53 rev. in 570,531.65d.	2 rev. in 59y.
Jupiter	225 rev. in 974,875.25d.	7 rev. in 83y.
Mars	923 rev. in 634,088.99d.	42 rev. in 79y.
Venus	1,199 rev. in 437,945.15d.	8 rev. in 8y.
	(750 synodic periods)	(5 synodic periods)
Mercury	46 rev. in 16,801.89d.	46 rev. in 46y.
	(145 synodic periods)	(145 synodic periods)

	Haridatta
Saturn	2 rev. − 0;1,15° in 58y. + 364d.
Jupiter	7 rev. − 0;38,55° in 82y. + 364d.
Mars	42 rev. + 0;33,43° in 78y. + 364d.
Venus	227 rev. + 0;18,26° in 226y. + 364d.
	(144 synodic periods)
Mercury	46 rev. − 2;8,39° in 45y. + 364d.
	(145 synodic periods)

The mean daily motions according to Haridatta are:

Saturn	0;2,0,16,49 °/d
Jupiter	0;4,59,10,43
Mars	0;31,26,31,16
Sun	0;59,8,10,12
Venus (anomaly)	0;37,30,51,20
Mercury (anomaly)	3;6,25,17,14
Moon	13;10,35,2,23
Lunar anomaly	13;3,53,58,36
Lunar node	− 0;3,10,48,25

Haridatta's epoch positions indicate the date 31 March 1638, which was a Saturday.

	Sidereal	Tropical	Sidereal-Tropical
Saturn	4,43;58,13°	5,6°	− 22
Jupiter	3,14;10,15	3,32	− 18
Mars	4,2;48,17	4,19	− 16
Sun	2;9,24	21	− 19
Venus	28;7,34	49	− 21
Mercury	5,40;37,15	5,55	− 14
Moon	5,14;20,22	5,40	− 26
Lunar node	4,22;52,40	4,41	− 18

Manuscripts:

1. Mars. 4869 ff. 1v–21.
2. Mercury. 4869 ff. 22–33.
3. Jupiter. 4869 ff. 34–47v.
4. Venus. 4869 ff. 48–85v.
5. Saturn. 4869 ff. 86–95v.
6. Lord of the year. Table of the lord of the year for 0 to 88 years. The parameter is 1;15,31,17,17 days per year; this means that one sidereal year in the *Jagadbhūṣaṇa* equals 6,5;15,31,17,17 days. The epoch value is 6;56,14,30—i.e., a Saturday.

Manuscripts: 4869 f. 96.

7. Epact. Table for 0 to 121 years. The yearly epact is taken to equal 11;3,53,22,40 tithis; the epoch value for Śaka 1560 is 26;11,41,50 tithis.

Manuscripts: 4869 ff. 96v–97.

8. Moon. Table for 0 to 121 years. The yearly mean motion of the Moon is taken to be 2,12;46,40,32°; the epoch value for Śaka 1560 is 5,14;20,22°.

Manuscripts: 4869 ff. 97–97v.

9. Anomaly. Table for 0 to 42 years. The yearly mean motion of the Anomaly is taken to be 1,32;6,8,48°; the epoch value for Śaka 1560 is 27;51,11°.

Manuscripts: 4869 ff. 97v–98.

10. Moon. Table of mean motion for 1 to 27 avadhis where an avadhi is defined as the time it takes the Sun to rise 15 times—i.e., 14 sunrise days. In other words, the tabulated values are mean longitudes of the Moon at sunrise of every 14th day. In order to compute these values Haridatta has added algebraically to the Moon's mean motion for 14 24-hour days (≈ 3,4;28,10°) the product of the Moon's mean daily motion (≈ 13;10,35°) multiplied by the sum of the equation of time and the equation of daylight (column 5 of table 16). The following table demonstrates this relationship.

avadhi	$\bar{\lambda}_{☾}$	text	Δ	computed correction
k = 1	0°	11^{s}29;54,17°	− 0;5,43°	− 0;5,55,45,45°
2	6^{s}4;28,10	6^{s}4;15,24	− 0;12,46	− 0;12,57,24,25
3	0^{s}8;56,20	0^{s}8;38,53	− 0;17,27	− 0;18,0,27,50
4	6^{s}13;24,30	6^{s}13;2,1	− 0;22,29	− 0;21,57,38,20
5	0^{s}17;52,40	0^{s}17;29,42	− 0;22,58	− 0;23,3,31,15
6	6^{s}22;20,50	6^{s}21;57,38	− 0;23,12	− 0;23,29,52,25
7	0^{s}26;49,0	0^{s}26;28,24	− 0;20,36	− 0;21,4,56,0
		7^{s}0;56,45, corr. to		
8	7^{s}1;17,10	7^{s}0;58,45	− 0;18,25	− 0;18,39,59,35
9	1^{s}5;45,20	1^{s}5;30,12	− 0;15,8	− 0;14,55,59,40
10	7^{s}10;13,30	7^{s}10;1,50	− 0;11,40	− 0;11,51,31,30
11	1^{s}14;41,40	1^{s}14;31,47	− 0;10,53	− 0;9,13,24,30
12	7^{s}19;9,50	7^{s}19;2,17	− 0;7,33	− 0;7,14,49,15
13	1^{s}23;38,0	1^{s}23;33,3	− 0;4,57	− 0;5,3,3,25
14	7^{s}28;6,10	7^{s}28;3,29	− 0;2,41	− 0;1,58,35,15
15	2^{s}2;34,20	2^{s}2;35,29	+ 0;1,9	+ 0;1,32,14,5
16	8^{s}7;2,30	8^{s}7;7,34	+ 0;5,4	+ 0;5,55,45,45
17	2^{s}11;30,40	2^{s}11;41,26	+ 0;10,46	+ 0;11,25,10,20
18	8^{s}15;58,50	8^{s}16;14,37	+ 0;15,47	+ 0;16,28,13,45

19	2^{s}20;27,0	2^{s}20;48,42	+0;21,42	+0;21,44,27,45
		8^{s}25;18,39, corr. to		
20	8^{s}24;55,10	8^{s}25;19,19	+0;24,9	+0;24,48,55,55
		2^{s}29;48,42, corr. to		
21	2^{s}29;23,20	2^{s}29;48,52	+0;25,32	+0;26,7,59,25
22	9^{s}3;51,30	9^{s}4;15,58	+0;24,28	+0;25,28,27,40
23	3^{s}8;19,40	3^{s}8;40,22	+0;20,42	+0;21,44,27,45
24	9^{s}12;47,50	9^{s}13;3,41	+0;15,51	+0;16,54,34,55
25	3^{s}17;16,0	3^{s}17;25,4	+0;9,4	+0;9,52,56,15
26	9^{s}21;44,10	9^{s}21;46,7	+0;1,57	+0;2,24,56,25
27	3^{s}26;12,20	3^{s}26;6,35	−0;5,45	−0;5,16,14,0

The differences between columns 3 and 4 are due to the fact that Haridatta used a slightly different value for the mean daily motion of the Moon than 13;10,35°.

Manuscripts: 4869 f. 98.

11. Anomaly. Table of mean motion for 1 to 27 avadhis, where an avadhi is defined as for table 10. I construct the same sort of table as before; $\lambda = (k-1)\ 3,2;54,36°$, whereas the entries in the last column equal 13;3,54° times the entries in column 5 of table 16. Note that the text entries for minutes and seconds for avadhis 19–24 have been lowered to avadhis 20–25.

avadhi	λ	text	Δ	computed correction
k = 1	0°	11^{s}29;54,20°	−0;5,40°	−0;5,52,45,18°
2	6^{s}2;54,36	6^{s}2;41,54	−0;12,42	−0;12,50,50,6
3	0^{s}5;49,12	0^{s}5;31,47	−0;17,25	−0;17,51,19,48
4	6^{s}8;43,48	6^{s}8;21,35	−0;22,13	−0;21,46,30,0
5	0^{s}11;38,24	0^{s}11;15,45	−0;22,39	−0;22,51,49,30
6	6^{s}14;33,0	6^{s}14;10,7	−0;22,53	−0;23,17,57,18
7	0^{s}17;27,36	0^{s}17;7,21	−0;20,15	−0;20,54,14,24
8	6^{s}20;22,12	6^{s}20;4,5	−0;18,7	−0;18,30,31,30
9	0^{s}23;16,48	0^{s}23;1,57	−0;14,51	−0;14,48,25,12
10	6^{s}26;11,24	6^{s}26;0,3	−0;11,21	−0;11,45,30,36
		0^{s}28;56,36, corr. to		
11	0^{s}29;6,0	0^{s}28;57,16	−0;8,44	−0;9,8,43,48
		7^{s}1;53,22, corr. to		
12	7^{s}2;0,36	7^{s}1;53,42	−0;6,34	−0;7,11,8,42
13	1^{s}4;55,12	1^{s}4;50,35	−0;4,37	−0;5,0,29,42
14	7^{s}7;49,48	7^{s}7;47,34	−0;2,14	−0;1,57,35,6
15	1^{s}10;44,24	1^{s}10;45,52	+0;1,28	+0;1,31,27,18
16	7^{s}13;39,0	7^{s}13;44,24	+0;5,24	+0;5,52,45,18
		1^{s}16;44,15, corr. to		
17	1^{s}16;33,36	1^{s}16;44,35	+0;10,59	+0;11,19,22,48
18	7^{s}19;28,12	7^{s}19;44,4	+0;15,52	+0;16,19,52,30
		1^{s}22;41,17, corr. to		
19	1^{s}22;22,48	1^{s}22;44,17	+0;21,29	+0;21,33,26,6
20	7^{s}25;17,24	7^{s}25;41,17	+0;23,53	+0;24,36,20,42
21	1^{s}28;12,0	1^{s}28;37,38	+0;25,38	+0;25,54,44,6
22	8^{s}1;6,36	8^{s}1;31,20	+0;24,44	+0;25,15,32,24
23	2^{s}4;1,12	2^{s}4;22,19	+0;21,7	+0;21,33,26,6
24	8^{s}6;55,48	8^{s}7;12,3	+0;16,15	+0;16,46,0,18
25	2^{s}9;50,24	2^{s}10;0,7	+0;9,43	+0;9,47,55,30
26	8^{s}12;45,0	8^{s}12;47,31	+0;2,31	+0;2,23,42,54
27	2^{s}15;39,36	2^{s}15;34,39	−0;4,57	−0;5,13,33,36

Manuscripts: 4869 f. 98.

12. Moon. Table of mean motions for 1 to 13 days. Mean daily motion equals 13;10,35°.

Manuscripts: 4869 f. 98.

13. Anomaly. Table of mean motion for 1 to 13 days. Mean daily motion equals 13;3,54°.

Manuscripts: 4869 f. 98v.

14. Moon. Table of mean motion for 1 to 60 ghaṭikās. Mean daily motion equals 13;10,35°. *Cf.* table 3 of the *Candrārkī*.

Manuscripts: 4869 f. 98v.

15. Moon. Table of lunar equations for arguments (kendrabhuja) of 0 to 30 (= 0° to 90°). This table has 31 lines and 5 columns. Column 1 gives the argument; when multiplied by 3 it gives the anomaly in degrees. Column 2 gives the equations; the maximum equation (at 90°) is 5;2,10°. Column 3 gives the differences between the entries in column 2. Column 4 gives the correction to be applied to the Moon's mean daily motion for each equation; the maximum correction is 1;8,49°, which gives a maximum daily motion of 14;19,24°, a minimum of 12,1,46°. Column 5 gives the differences between the entries in column 4.

Manuscripts: 4869 f. 99.

16. Table for transforming avadhis of 14 24-hour days into avadhis of 14 sunrise days. Two components must be known: the equation of time, which is the difference between 24 hours and the interval between two consecutive transits by the Sun of the six o'clock circle; and the equation of daylight, which is the difference between the Sun's transit of the six o'clock circle and sunrise at a given latitude ϕ. Furthermore, the equation of time has two components: the first is due to the change in solar velocity, and the second to the obliquity of the ecliptic.

Table 16 has 5 columns. The first contains the arguments: 1 to 27 avadhis. Column 2 contains that component of the equation of time which is due to the change in solar velocity; it is called yātaphala. Column 3 contains the equation of daylight; it is called cara. Column 4 contains the second component of the equation of time; it is called bhujaphala. And column 5 is the sum of columns 2 to 4; it is called trayaikya.

The entries in column 3, it should be noted, are equal to half the differences between the lengths of daylight and 12 hours. The lengths of daylight are tabulated in table 18. From the scholium on that table we learn that the longest daylight according to chapter 1 of the *Jagadbhūṣaṇa* is 33;48 ghaṭikās or $13^h31^m12^s$. This means that the latitude ϕ for which the tables were computed is approximately 24°N—i.e., the latitude of Ujjain.

The entries in the table are in ghaṭikās or sixtieths of a day.

1	2	3	4	5	2+4 (equation of time)
k = 1	+0;14	+0;34	−0;21	+0;27	−0;7
		+1;6, corr. to		+1;17, corr. to	

2	+0;22	+0;56	-0;19	+0;59	+0;3
3	+0;24	+1;14	-0;16	+1;22	+0;8
4	+0;22	+1;31	-0;13	+1;40	+0;9
	+0;34, corr. to				
5	+0;14	+1;39	-0;8	+1;45	+0;6
6	+0;3	+1;47	-0;3	+1;47	0;0
7	-0;9	+1;43	+0;2	+1;36	-0;7
8	-0;18	+1;36	+0;7	+1;25	-0;11
9	-0;24	+1;21	+0;11	+1;8	-0;13
10	-0;24	+1;3	+0;15	+0;54	-0;9
				+0;52, corr. to	
11	-0;19	+0;43	+0;18	+0;42	-0;1
12	-0;9	+0;21	+0;21	+0;33	+0;12
13	+0;3	-0;2	+0;22	+0;23	+0;25
14	+0;14	-0;26	+0;21	+0;9	+0;35
15	+0;22	-0;49	+0;20	-0;7	+0;42
		-6;9, corr. to			
16	+0;25	-1;9	+0;17	-0;27	+0;42
17	+0;22	-1;28	+0;14	-0;52	+0;36
18	+0;14	-1;38	+0;9	-1;15	+0;23
19	+0;3	-1;46	+0;4	-1;39	+0;7
20	-0;8	-1;43	-0;2	-1;53	-0;10
21	-0;18	-1;34	-0;7	-1;59	-0;25
				-1;26, corr. to	
22	-0;24	-1;20	-0;12	-1;56	-0;36
		-0;9, corr. to		-1;27, corr. to	
23	-0;24	-0;59	-0;16	-1;39	-0;40
24	-0;19	-0;39	-0;19	-1;17	-0;38
				-1;17, corr. to	
25	-0;9	-0;15	-0;21	-0;45	-0;30
	+0;12, corr. to	+0;39, corr. to		-0;1, corr. to	
26	+0;2	+0;9	-0;22	-0;11	-0;20
				+0;28, corr. to	
27	+0;13	+0;32	-0;21	+0;24	-0;8

Manuscripts: 4869 ff. 99–99v.

17. Sun. Table of true longitudes for 1 to 27 avadhis = 0 to 364 days. This table has 27 lines and 4 columns. Column 1 gives the argument in days and avadhis. Column 2 gives the true longitudes of the Sun. Column 3 gives the daily motion of the Sun at that longitude. Column 4 gives the differences between the entries in column 3. The maximum daily motion of the Sun ≈ 1;1,23°; the minimum daily motion ≈ 0;56,54°. The mean motion, then, is approximately 0;59,8° as it should be.

In the following table I have computed $\bar{\lambda}$ with that mean daily motion—0;59,8°. It emerges from the table that the apogee is at about Gemini 20°, and that the maximum equation is about 2;13°.

avadhis	$\bar{\lambda}_{\odot}$	text	equation	daily motion
k = 1	0°	0^{s}2;9,24°	+2;9,24°	0;58,40°
2	0^{s}13;47,52	0^{s}15;46,17	+1;58,25	0;58,10
3	0^{s}27;35,44	0^{s}29;16,55	+1;41,11	0;57,42
4	1^{s}11;23,36	1^{s}12;41,39	+1;18,3	0;57,20
5	1^{s}25;11,28	1^{s}26;2,19	+0;50,51	0;57,4
6	2^{s}8;59,20	2^{s}9;20,9	+0;20,49	0;56,55
7	2^{s}22;47,12	2^{s}22;36,50	-0;10,22	0;56,54
8	3^{s}6;35,4	3^{s}5;54,16	-0;40,48	0;57,1
9	3^{s}20;22,56	3^{s}19;14,6	-1;8,50	0;57,35 (*sic!*)
10	4^{s}4;10,48	4^{s}2;37,59	-1;32,49	0;57,35
11	4^{s}17;58,40	4^{s}16;7,11	-1;51,29	0;57,58
12	5^{s}1;46,32	4^{s}29;43,0	-2;3,32	0;58,30
13	5^{s}15;34,24	5^{s}13;26,0	-2;8,24	0;59,1
14	5^{s}29;22,16	5^{s}27;16,30	-2;5,46	0;59,36
15	6^{s}13;10,8	6^{s}11;14,41	-1;55,27	1;0,6
16	6^{s}26;58,0	6^{s}25;19,31	-1;40,29	1;0,32
17	7^{s}10;45,52	7^{s}9;30,22	-1;15,30	1;0,55
18	7^{s}24;33,44	7^{s}23;45,46	-0;47,58	1;1,11
19	8^{s}8;21,36	8^{s}8;4,7	-0;17,29	1;1,21
20	8^{s}22;9,28	8^{s}22;23,32	+0;14,4	1;1,23
21	9^{s}5;57,20	9^{s}6;42,21	+0;45,1	1;1,16
22	9^{s}19;45,12	9^{s}20;58,40	+1;13,28	1;1,3
23	10^{s}3;33,4	10^{s}5;10,43	+1;37,39	1;0,43
24	10^{s}17;20,56	10^{s}19;17,21	+1;56,25	1;0,17
25	11^{s}1;8,48	11^{s}3;17,22	+2;8,34	0;59,46
26	11^{s}14;56,40	11^{s}17;10,8	+2;13,28	0;59,16
27	11^{s}28;44,32	0^{s}0;55,21	+2;10,49	0;58,42

Manuscripts: 4869 f. 99v.

18. Length of daylight. Table of ghaṭikās of daylight at the beginnings of 1 to 27 avadhis. Though corrupt, the text can be restored with the help of column 3 of table 16.

avadhi	ghaṭikās of daylight
k = 1	31;8
2	31;52
3	32;28
4	33;2
5	33;18
6	33;34
7	33;26
8	33;12
9	32;6 (read 32;42)
10	32;6
11	32;6 (read 31;26)
12	31;36 (read 30;42)
13	32;42 (read 29;56)
14	29;56 (read 29;8)
15	29;8 (read 28;22)
16	28;22 (read 27;42)
17	27;42 (read 27;4)
18	27;4 (read 26;44)
19	26;44 (read 26;28)
20	26;28 (read 26;34)
21	26;53 (read 26;52)
22	26;53 (read 27;20)
23	27;20 (read 28;2)
24	28;2 (read 28;42)
25	28;42 (read 29;30)
26	29;30 (read 30;18)
27	30;18 (read 31;4)
	31;4 in margin

Manuscripts: 4869 f. 99v.

19. The lunar node. Table of mean motion for 0 to 92 years. The mean yearly motion is -19;21,34°—a number which is recorded in the margin of 4869. Below each yearly entry is dutifully recorded the mean daily

motion: −0;3,11°. The epoch position is 4,22;52,40°. Manuscripts: 4869 f. 100; 4944 ff. 151–151v; *cf.* 4946 XXXIV f. 138.

20. The lunar node. Table of mean motion for 1 to 27 avadhis. The mean yearly motion, as above, is −19;21, 34°.

Manuscripts: 4869 f. 100v; 4944 f. 151; 4946 XXXIV f. 138v.

21. Table of the day and fraction of a day on which the Sun enters each of the 12 zodiacal signs and 27 nakṣatras. The first column gives the day-number such that, when divided by 7, the same week-day will result as is given in line 3. Column 2 names the sign or nakṣatra in an abbreviated manner. Column 3 gives the week-day of the Sun's entry, beginning with 4;48,21 (Thursday) for the entry into Aries and Aśvinī. The last entry merely repeats the first and must be ignored.

1	2	3
362	Meṣa	4;48,21
362	Aśvinī	4;48,21
11	Bharaṇī	4;30,26
25	Kṛttikā	4;18,49
29 [28]	Vṛṣa	0;46,49
39	Rohiṇī	4;12,44
53	Mṛgaśiras	4;10,47
60	Mithuna	4;11,2
67	Ārdrā	4;11,48
81	Punarvasu	4;14,1
92 [91]	Karkaṭa	0;45,32
95	Puṣya	4;15,53
109	Āśleṣā	4;15,48 [4;16,48]
123	Siṃha	4;16,23 [4;15,23]
123	Maghā	4;16,23 [4;15,23]
137	Pūrvaphālgunī	4;4,9 [4;9,9]
150	Uttaraphālgunī	3;50,2 [3;55,2]
154	Kanyā	0;50,32 [0;15,32]
164	Hasta	[3;30,]18
178	Citra	3;0,29
184	Tulā	2;44,50
191	Svāti	2;26,27
204	Viśākhā	1;44,38
214	Vṛścika	4;39,15
217	Anurādhā	0;46,46
236 [231]	Jyeṣṭhā	0;3,52
244	Dhanuḥ	6;7,12
244	Mūla	6;7,12
257	Pūrvāṣāḍhā	5;8,25
270	Uttarāṣāḍhā	4;9,15
273	Makara	0;24,34
283	Śravaṇa	3;11,15
296	Dhaniṣṭhā	2;16,1
303 [302]	Kumbha	1;50,0
309	Śatabhiṣak	1;25,9
332 [322]	Pūrvabhādrapadā	0;39,40
332	Mīna	3;39,52
336	Uttarabhādrapadā	0;0,44
349	Revatī	6;29,7
362	Meṣa	4;48,21

Manuscripts: 4869 f. 100v.

ANONYMOUS OF 1638

I. Manuscripts:

*Poleman 4942 (Smith Indic 119) f. 18.

II. Tables:

1. Table of tithikendrakṣepakas for 1 to 13 synodic months, and of cālakas measured in palas. The entries for tithikendrakṣepakas increase from 0;10,38, at 1 to 24;16,14, at 13. See table 3 of the Anonymous of 1790.

Manuscripts: 4942 f. 18.

2. Table of nakṣatrakendrakṣepakas for 1 to 14 sidereal months. The entries decrease from 27;0,0 at 1 to 24;14, 18 at 14. See table 4 of the Anonymous of 1790.

Manuscripts: 4942 f. 18.

3. Table of tithikendrakṣepakas measured in days (modulo 7) for 1 to 13 synodic months, and of cālakas measured in ghaṭikās. The length of a mean synodic month is taken to be 29;31,54 days; the epoch position is 6;11,48 days. *Cf.* table 3 of the Anonymous of 1790.

Manuscripts: 4942 f. 18.

4. Table of nakṣatrakendrakṣepakas measured in days (modulo 7) for 1 to 14 sidereal months, and of cālakas measured in palas. The length of a mean sidereal month is taken to be 27;19,22 days; the epoch position is 6;1,15 days. *Cf.* table 4 of the Anonymous of 1790.

Manuscripts: 4942 f. 18.

5. Table of intercalary months for Śaka 1560 to 1690 (A.D. 1638 to 1768). This overlaps with table 2 of the Anonymous II of 1578 somewhat, but does not entirely agree with it.

1560	Śrāvaṇa	1612	Vaiśākha
1563	Jyeṣṭha	1614	Bhādrapada
1566	Caitra	1617	Āṣāḍha
1568	Śrāvaṇa	1620	Jyeṣṭha
1572	Āṣāḍha	1622	Āśvina
1574	Vaiśākha	1625	Śrāvaṇa
1576	Bhādrapada	1628	Āṣāḍha
1579	Śrāvaṇa	1631	Vaiśākha
1582	Jyeṣṭha	1633	Bhādrapada
1585	Caitra	1636	Āṣāḍha
1587	Śrāvaṇa	1639	Jyeṣṭha
1590	Āṣāḍha	1641	Āśvina
1593	Vaiśākha	1644	Śrāvaṇa
1595	Bhādrapada	1647	Āṣāḍha
1598	Śrāvaṇa	1650	Jyeṣṭha
1601	Jyeṣṭha	1652	Bhādrapada
1603	Āśvina	1655	Āṣāḍha
1606	Śrāvaṇa	1658	Jyeṣṭha
1609	Āṣāḍha	1660	Bhādrapada

1663 Śrāvaṇa	1680 Āśvina
1666 Āṣāḍha	1682 Śrāvaṇa
1669 Caitra	1685 Āṣāḍha
1671 Bhādrapada	1688 Vaiśākha (in margin)
1674 Śrāvaṇa	1690 Bhādrapada
1677 Jyeṣṭha	

Manuscripts: 4942 f. 18.

6. Table of yogakendrakṣepakas for 1 to 15 periods (see table 7), and of cālakas measured in palas. The entries decrease from 29;6,27 at 1 to 26;38,3 at 15 modulo c. 30. See table 5 of the Anonymous of 1790.

Manuscripts: 4942 f. 18v.

7. Table of yogakendrakṣepakas measured in days (modulo 7) for 1 to 15 periods, and of cālakas measured in ghaṭikās. The period is 25;25,10 days; the epoch position is 5;52,8 days.

Manuscripts: 4942 f. 18v.

8. Table of week-days on which the Sun enters each of the 27 nakṣatras, and of the Sun's progress on that day.

Aśvinī	4;46,58d	0;58,33°
Bharaṇī	4;28,10	0;57,28
Kṛttikā	4;16,6	0;57,23
Rohiṇī	4;10,4	0;57,13
Mṛgaśiras	4;8,52	0;57,3
Ārdrā	4;10,52	0;56,50
Punarvasu	4;14,10	0;56,54
Puṣya	4;17,23	0;57,5
Āśleṣā	4;18,12	0;57,14
Maghā	4;15,3	0;57,42
Pūrvaphālgunī	4;6,38	0;58,10
Uttaraphālgunī	3;51,43	0;58,41
Hasta	3;29,52	0;59,10
Citra	3;0,58	0;59,44
Svāti	2;24,59	1;0,10
Viśākhā	1;42,31	1;0,35
Anurādhā	0;54,27	1;0,58
Jyeṣṭhā	0;1,45	1;1,16
Mūla	6;5,53	1;1,21
Pūrvāṣāḍhā	5;8,3	1;1,22
Uttarāṣāḍhā	4;10,2	1;1,19
Śravaṇa	3;12,45	1;1,8
Dhaniṣṭhā	2;18,2	1;1,1
Śatabhiṣak	1;27,11	1;0,26
Pūrvabhādrapadā	0;41,23	1;0,18
Uttarabhādrapadā	0;1,44	0;59,27
Revatī	6;28,28	0;58,58

Manuscripts: 4942 f. 18v.

9. Table of week-days on which the Sun enters each of the zodiacal signs, and of the Sun's progress on that day. The week-days are identical with those given in table 3 of the Anonymous of 1594, but the Sun's progresses differ:

Aries	0;58,28°	Leo	0;57,47°	Sagittarius	1;1,11°
Taurus	0;57,28	Virgo	0;58,8	Capricorn	1;1,17
Gemini	0;57,33	Libra	1;0,0	Aquarius	1;0,27
Cancer	0;57,33	Scorpio	1;1,7	Pisces	0;59,7

Manuscripts: 4942 f. 18v.

THE *PAÑCĀṄGAVIDYĀDHARĪ* OF VIDYĀDHARA

I. Life of Vidyādhara:

Our information about Vidyādhara comes mainly from the first 4 verses of his *Pañcāṅgavidyādharī.*

viditaparamatattvā jñānino yaṃ smaranti
suramanujamanuṣyāḥ siddhideyaṃ namanti /
sakalabhuvananātho rādhayā suṃyutātmā
diśatu ⟨sa⟩ śamasiddhiṃ vāci (?) dāmodarasya // 1 //
puravibhūṣaṇajīrṇagaḍhe pure 'calavibhūṣaṇaraivatikācale /
sakalasaccarite kuladīpako hariharo 'tra vasiṣṭhakule prabhūt // 2 //
tasyādyaputro giriśabdapūrvo nārāyaṇajñātivibhūṣaṇo 'sti/
nārāyaṇākṣo 'dya tadādyasūnur gārgeyavidyādhara eva jātaḥ // 3 //
so 'ham mate brahmabhave tathārthe pañcāṅgavidyāḍhyadharim pravakṣye/
śake śarāṅgākṣakubhir vihīne gatābdanighnā guṇakā dhruvāḥ syuḥ // 4 //

Thus Vidyādhara was the son of Nārāyaṇa and the grandson of Harihara of the Girinārāyaṇa branch of the Vasiṣṭhagotra, who lived in Jīrnagaḍha on Mount Raivatika (Junāgaḍh on Mount Girnar in Saurāṣṭra). He wrote the *Grahavidyādharī* in Śaka 1560 (A.D. 1638); it is preserved in IO 2961 (2083c) 6 ff. The epoch of the *Pañcāṅgavidyādharī* is Śaka 1565 (A.D. 1643). The text informs us (4778 f. 3) that it was written during the reign of Vīrabhadra, the king of Rājakoṭa (Rājkot in Saurāṣṭra).

II. Manuscripts:

IO 2960 (2529b) 6 ff. A.D. 1726.
*Poleman 4778 (Smith Indic 30) 18 ff. Saṃ. 1806 = A.D. 1749.

III. Tables:

1. Table of yearly parameters (guṇakas) and epoch positions (kṣepakas), for the Epact, the Lord of the Year, and the "mandatihi" according to the school of the *Brahmatulya* (brahmapakṣa) and that of the *Āryabhaṭīya* (āryapakṣa).

According to the brahmapakṣa:

	yearly parameters	epoch positions for Śaka 1565
Epact	11;3,53,22,40 tithis	22;32,23,57,16 tithis
Lord of Year	1;15,31,17,17 days	6;13,52,40,20 days
Mandatithi	7;40,30,46,29 tithis	10;45,37,5,50 tithis

	Rāmabījas
Tithibhoga	+1,15
Tithikendra	-1,30
Nakṣatrabhoga	+1,0
Nakṣatrakendra	-2,30
Yogabhoga	+0,45
Yogakendra	-2,0

According to the āryapakṣa:

	yearly parameters	epoch positions for Śaka 1565
Epact	11;3,52,53,52 tithis	22;24,36,45,0 tithis
Lord of Year	1;15,31,15,0 days	6;10,50,45,0 days
Mandatithi	7;40,29,42,12 tithis	10;33,55,1,30 tithis

Cf. table 8 (of the *Tithicintāmaṇi*).

Manuscripts: 4778 f. 3.

2. Table of the tithis and week-days on which the Sun enters each of the 12 zodiacal signs, and of the lengths of daylight on those days. The week-days are identical with those in table 3 of the Anonymous of 1594 and table 9 of the Anonymous of 1638. The maximum length of daylight is 33;21,30 ghaṭikās when the Sun is at Gemini 0°.

Manuscripts: 4778 f. 3.

3. Table of yearly parameters:

Epact	11;3,53
Lord of Year	1;15,31
Mandatithi	7;40,31
Tithidhruva	10;56,7
Nakṣatradhruva	10;2,30
Yogadhruva	10;2,30
Tithibhogya	1;11,42
Nakṣatrabhogya	1;18,4
Yogabhogya	1;17,53
Tithikendra	7;5,49
Nakṣatrakendra	7;0,27
Yogakendra	7;31,45

Manuscripts: 4778 f. 3v.

4. Table of tithis and week-days on which the Sun enters each of the 27 nakṣatras, and of the lengths of daylight on those days. The week-days are identical with those in table 8 of the Anonymous of 1638. The maximum length of daylight is 33;29 ghaṭikās when the Sun enters Ārdrā, the minimum 26;28 ghaṭikās when it enters Pūrvāṣāḍhā.

Manuscripts: 4778 f. 4.

5. Table of tithikendras (modulo 27;59,33) and of week-days with their respective cālakas for 0 to 37 periods of 10 days. This is table 11 of the Anonymous of 1741.

Manuscripts: 4778 ff. 4v–5.

6. Table of nakṣatrakendras (modulo 27;13,48) and of week-days with their respective cālakas for 0 to 41 periods of 9 days. This is table 12 of the Anonymous of 1741.

Manuscripts: 4778 ff. 5v–6.

7. Table of yogakendras (modulo 29;16,1) and of week-days with their respective cālakas for 0 to 43 periods of 9 days. This is table 13 of the Anonymous of 1741.

Manuscripts: 4778 ff. 6v–7.

8. Table 5 of the Anonymous of 1594.

Manuscripts: 4778 f. 7v.

9. Table of corrections to the tithikendra for 0 to 27 days horizontal, and for 0 to 59 ghaṭikās vertical. The maximum correction is 24;47; *cf.* table 16 of the Anonymous of 1741.

Manuscripts: 4778 ff. 8v–12.

10. Table of corrections to the nakṣatrakendra for 0 to 27 days horizontal, and for 0 to 59 ghaṭikās vertical. The maximum correction is 22;56; *cf.* table 17 of the Anonymous of 1741.

Manuscripts: 4778 ff. 12v–15.

11. Table of corrections to the yogakendra for 0 to 29 days horizontal, and for 0 to 59 ghaṭikās vertical. The maximum equation is 21;20; *cf.* table 18 of the Anonymous of 1741.

Manuscripts: 4778 ff. 15v–18v.

THE *KHECARADĪPIKĀ* OF KALYĀṆA

I. The Life of Kalyāṇa:

Our only information concerning Kalyāṇa is derived from the first three verses of his work:

śrīmantaṃ gaṇanāyakaṃ ravividhukṣonījacandrātmajān
vāṇīnāthakavidineśajatamaḥketūn sindhiṃ bhāratīm /
natvā sadgurupādapaṅkajam ahaṃ vakṣye vidajñānatā
hantrīṃ khecaradīpikāṃ vitanute kalyāṇanāmā kaviḥ
// 1 //
brahmeśāryabhaṭārkabhāskaramunipramukhyaiḥ parākarmabhiḥ
ṣaḍbhir bhuktipuraḥsaraiḥ khalu kṛtā mandāśupātair vinā /
tatsārād avagamya sāram adhunā khecāriṇāṃ sāriṇī
śiṣyaprītikṛtāvadher mukhagatā rāśyādikātrāpi ca //2//
śāko bhūhayabāṇacandrarahito śeṣaṃ gatābdās tu te
guṇāḥ svair guṇakais tato nijanijakṣepair yutāḥ syuḥ sphuṭāḥ /
śuddhyabdau tithivarajau ca khecarān rāśyādikān kārayet
te sarve 'bdamukhā bhavanti nitarāṃ karmopayogyāḥ same // 3 //

In fact, all this tells us is that the epoch of the work is Śaka 1571 (A.D. 1649).

II. Manuscripts:

*Poleman 4950 (Smith Indic 17) 12 ff

III. Tables:

1. Table of yearly mean motions (taken, where possible, from the *Mahādevī*) and of epoch positions for the Epact, Lord of the Year, Moon (2,12;46,40,32° per year), Lunar Anomaly (1,32;6,8,48° per year), Mars, Mercury's Anomaly, Jupiter, Venus' Anomaly, Saturn, and the Lunar Node (−19;21,33,59° per year). The epoch positions indicate as date Saturday, 31 March 1649 in the Julian calendar.

	Sidereal	Tropical	Sidereal-Tropical
Epact	27;55,43,59,20 tithis		
Lord of the Year	6;46,58,39,38 days		
Mars	2,42;43,13°	147°	+15°!
Mercury's anomaly	2,56;44,15	(Mercury) 22	
Jupiter	2,47;23,13	183	−16
Venus' anomaly	2,6;3,23	(Venus) 66	
Saturn	58;7,20	72	−14
Lunar Node	49;55,27	c.65	

Manuscripts: 4950 f. 1v.

2. Sun. Table of solar longitudes for 1 to 27 avadhis. This table has 4 columns, listing respectively the avadhi-number, the longitude of the Sun, the daily progress of the Sun at that longitude, and the differences between successive entries in column 3. Its arrangement, therefore, is identical with that of table 17 in the *Jagadbhūṣaṇa*. Column 2 is the only one to show divergence in its entries; and this is due to the fact that the initial longitudes—2;9,24° in the *Jagadbhūṣaṇa* and 2;9,13° in the *Khecaradīpikā*—differ. The value 2;9,13° is from table 1 of the *Candrārkī*.

Manuscripts: 4950 f. 2v.

3–7. Planets. These tables, 12 for each planet, are derived directly from the *Mahādevī*. Each of the 12 tables for each planet begins with a situation wherein the mean Sun is at Aries 0° (sidereal), and the mean planet (or its anomaly, in the case of Venus and Mercury) is at the beginning of one of the zodiacal signs. Each table gives the *Mahādevī's* columns A,B,D, and F, with G in the margin.

The columns B are copied directly from the *Mahādevī* for N = 0,5,10,15,20,25,30,35,40,45,50, and 55. The columns D are copied directly from the *Mahādevī* for N = 2,7,12,17,22,27,32,37,42,47,52, and 57. Column F is computed by using the formula:

$$F(n) = \frac{800}{5\Delta C}$$

where $\Delta C = C(N_n) - C(N_{n+5})$ of the *Mahādevī* tables. G is copied directly from the *Mahādevī*.

Manuscripts:

3. Mars. 4950 ff. 3–4v.
4. Mercury. 4950 ff. 5–7v.
5. Jupiter. 4950 ff. 8–9v.
6. Venus. 4950 ff. 10–11v.
7. Saturn. 4950 ff. 12–12v.

ANONYMOUS OF 1656

I. Manuscripts:

*Poleman 4946 (Smith Indic MB) XXI ff. 2 and 8, and an unnumbered f.
*Poleman 4946 (Smith Indic MB) LXVI ff. 1–2.
*Poleman 4946 (Smith Indic MB) LXXXVI f. 1.

II. Tables:

1. Tables of mean motions of the five star-planets in periods of 10 years from Śaka 1578 to 2018 (A.D. 1656 to 2096). To convert these mean motions into degrees one must multiply them by 3; therefore, they presuppose the existence of tables of true longitudes similar to those of the *Mahādevī*, but computed for increments of 3° in a planet's mean longitude instead of 6°. The parameters are identical with those of Mahādeva, as can be shown by computing the mean motions for 30 years which are precisely the *Mahādevī's* mean motions for 60 years.

	mean motions for 10 years	mean motions for 30 years = *Mahādevī's* mean motions for 60 years
Mars	38;0,31,7,0	⟨3⟩1,54;1,33,21
Mercury	⟨1⟩,2;31,3,27,20	⟨9⟩,7;33,10,22
Jupiter	⟨1⟩,41;10,21,42,20	5,3;31,5,7
Venus	30;39,39,22,0	⟨37⟩,31;58,58,6
Saturn	40;42,51,24,0	2,2;8,34,12

In computing his epoch positions, however, the author of this set of tables made a mistake and did not double the *Mahādevī's* mean positions for the beginning of Śaka 1578.

	Epoch positions = those computed from *Mahādevī* for Śaka 1578.
Mars	10;25,21,6,48
Mercury	33;19,52,49,22
Jupiter	3;18,29,7,36
Venus	43;44,14,54,48
Saturn	23;56,12,53,36

Manuscripts: 4946 XXI ff. 2–2v.

2. Table of the mean positions of the Moon and of the Lunar Anomaly for Śaka 1601 to 1685 (A.D. 1679 to 1763). The mean yearly motions are respectively 2,12; 46,40,20° and 1,32;6,8°; the epoch positions for Śaka 1601 (A.D. 1679) are 5,58;14,4° and 3,24;3,12°.

Also noted in the manuscript are the intercalary months for Śaka 1601 to 1658 (A.D. 1679 to 1736):

1601 Jyeṣṭha	1609 Āṣāḍha	1617 Āṣāḍha
1603 Āṣāḍha	1612 Vaiśākha	1620 Jyeṣṭha
1606 Āṣāḍha	1614 Māgha	1622 Āṣāḍha

1625 Āṣāḍha	1644 Āṣāḍha
1628 Āṣāḍha	1647 Āṣāḍha
1631 Vaiśākha	1650 Vaiśākha
1633 Māgha	1652 Māgha
1636 Āṣāḍha	⟨1655 Āṣāḍha⟩
1639 Jyeṣṭha	1658 Jyeṣṭha
1641 (?) Āṣāḍha	

Manuscripts: 4946 XXI ff. 8–8v.

3. Table relating to length of daylight; its precise structure remains obscure.

Manuscripts: 4946 XXI unnumbered f.; 4946 LXVI ff. 1–2; 4946 LXXXVI ff. 1–1v.

THE *GRAHAPRABODHASĀRIṆĪ* OF YĀDAVA

I. The Life of Yādava:

This work is based on the *Grahaprabodha* of Nāgeśa (or Nāganātha), the son of Śiva and grandson of Keśava, who wrote in Śaka 1541 (A.D. 1619); Yādava was Nāgeśa's pupil, as we learn from the opening verse of his udāharaṇa:

nāganāthagurum natvā yādavena vitanvate /
grahaprabhūtisaṃbhūtim udāharaṇam ādarāt // 1 //

Yādava's date is further pinpointed by the example which he uses in this udāharaṇa, a nativity dated Wednesday, the 5th tithi of the kṛṣṇapakṣa of Mārgaśīrṣa in Śaka 1585; this corresponds to 9 December 1663 in the Julian calendar. Nāgeśa, it is worth noting, lived in the Khecaramaṇḍala; presumably this is the Kheṭakaviṣaya in Gujarat.

II. Manuscripts:

DC Gorhe 76 13 ff. Śaka 1685 = A.D. 1763.
*Poleman 4777 (Smith Indic 33) 33 ff. Śaka 1793 = A.D. 1871.
Vārāṇasī 35648 ff. 1–38. Saṃ. 1939 = A.D. 1882. Identical with Benares (1903) 1226 39 ff. Saṃ. 1839 (*sic!*) = A.D. 1782.
ASBombay 233 11 ff.
Baroda 3108 ff. 4–29.
CP, Hiralal 1539 (from Kārañjā in Akolā District).
Vārāṇasī 35647 ff. 1–4.

III. Tables:

1. Tables of mean motions for 7 and for 14 days, for the Sun, the Moon, the Lunar Apogee, Mars, Mercury's anomaly, Jupiter, Venus' anomaly, Saturn, and the Lunar Node.

Manuscripts: 4777(a) f. 1v.

2. Tables of mean motions (madhyapattra) of the Sun for 0 to 9 days, 1 (×10) to 9 (×10) days, 1 (×100) to 9 (×100) days, and 1 (×1000) to 4 (×1000 days); the longitude for 0 days is the epoch position. A second table of mean motions (cakrāñka) uses as its basic period 11 years (4016 days; *cf.* the Anonymous of 1520); mean motions are tabulated here for 0 to 9, 1 (×10) to 9 (×10), 1 (×100) to 9 (×100), and 1 (×1000) to 4 (×1000) of these periods. The mean motion tables of all the planets are arranged in the same way.

The epoch positions indicate as date 5 March 1619 in the Julian calendar.

	Sidereal	Tropical	Sidereal-Tropical
Saturn	54;5,0,23°	63°	− 9°
Jupiter	5,35;38,59,9	351	−16
Mars	2,47;51,26,31	188	−20
Sun	5,34;17,29,10	354	−20
Venus (anom.)	3,14;27,59,40	(Venus) 327	
Mercury (anom.)	4,1;36,24,8	(Mercury) 327	
Moon	5,28;20,55,52	350	−22
Lunar Apogee	3,52;54,40,51		
Lunar Node	4,32;4,49,12	c290	

The mean daily motions upon which the tables are based are:

Saturn	0;2,0,23,5°
Jupiter	0;4,59,8,34
Mars	0;31,26,31
Sun	0;59,8,10,18
Venus (anom.)	0;36,59,40,6
Mercury (anom.)	3;6,24,8,6
Moon	13;10,34,51
Lunar Apogee	0;6,40,51,24
Lunar Node	−0;3,10,48,30

Manuscripts: 4777(a) ff. 2–2v.

3. Mean motion tables for the Moon.
Manuscripts: 4777(a) ff. 3–3v.

4. Mean motion tables for the Lunar Apogee.
Manuscripts: 4777(a) ff. 4–4v.

5. Mean motion tables for Mars.
Manuscripts: 4777(a) ff. 5–5v.

6. Mean motion tables for Mercury's anomaly.
Manuscripts: 4777(a) ff. 6–6v.

7. Mean motion tables for Jupiter.
Manuscripts: 4777(a) ff. 7–7v.

8. Mean motion tables for Venus' anomaly.
Manuscripts: 4777(a) ff. 8–8v.

9. Mean motion tables for Saturn.
Manuscripts: 4777(a) ff. 9–9v.

10. Mean motion tables for the Lunar Node.
Manuscripts: 4777(a) ff. 10–10v.

11. Table of the equation of the center of the Sun. There are 6 columns. In column 1 are listed the degrees

of argument from 0° to 90°; in column 2 the equations to seconds; in column 3 the guṇakas; in column 4 the hāras; in column 5 the increments to (or decreases from) the mean daily motions; and in column 6 the hāras for that. The maximum equation is 2;10,45°; the maximum increment 0;2,15°.

Manuscripts: 4777(a) ff. 11–12v.

12. Table of the equation of the center of the Moon, set up as is that for the Sun. The maximum equation is 5;1,41°; the maximum increment 1;8,15°.

Manuscripts: 4777(a) ff. 13–14v.

13. Table of Mars' equation of the conjunction. There are two columns, one of the argument (0° to 180°) and one of the equations. In the margin are listed the guṇas and daily progresses for every 15° of argument. All the planets' tables of equations of the conjunction are set up in this way, as are also their tables of equations of the center, save that the argument in the latter is only from 0° to 90°. The maximum equation of the conjunction for Mars is 40;0,0° at 135°.

Manuscripts: 4777(a) ff. 15–16v.

14. Table of Mercury's equation of the conjunction. The maximum equation is 21;12,0° at 105° to 120°.

Manuscripts: 4777(a) ff. 17–18v.

15. Table of Jupiter's equation of the conjunction. The maximum equation is 10;48,0° at 105°.

Manuscripts: 4777(a) ff. 19–20v.

16. Table of Venus' equation of the conjunction. The maximum equation is 46;6,0° at 135°.

Manuscripts: 4777(a) ff. 21–22v.

17. Table of Saturn's equation of the conjunction. The maximum equation is 5;42,0° at 90° to 105°.

Manuscripts: 4777(a) ff. 23–24v.

18. Table of Mars' equation of the center. The maximum equation is 13;0,0°.

Manuscripts: 4777(a) ff. 25–25v.

19. Table of Mercury's equation of the center. The maximum equation is 3;36,0°.

Manuscripts: 4777(a) ff. 26–26v.

20. Table of Jupiter's equation of the center. The maximum equation is 5;42,0°.

Manuscripts: 4777(a) ff. 27–27v.

21. Table of Venus' equation of the center. The maximum equation is 1;30,0°.

Manuscripts: 4777(a) ff. 28–28v.

22. Table of Saturn's equation of the center. The maximum equation is 9;18,0°.

Manuscripts: 4777(a) ff. 29–29v.

23. Table of guṇas and gatiphalas for Mars, Venus, and Mercury for anomalies of 165° to 180°.

Manuscripts: 4777(a) f. 30.

24. Table of mean daily motions of the Sun, the Moon, the Lunar Apogee, Mars, Mercury's anomaly, Jupiter, Venus' anomaly, Saturn, Rāhu, and Ketu in degrees, minutes, seconds, and thirds, and again in minutes and seconds. *Cf.* table 26a (of the *Makaranda*).

Manuscripts: 4777(a) f. 30v.

ANONYMOUS OF 1704

I. Manuscripts:

*Poleman 4766 (Smith Indic 139) 138 ff.
*Poleman 4805, 4821, 4822, 4884, 4904, 4916 (Smith Indic 100) ff. 3–22.
*Poleman 4865 (Smith Indic 85) 2 ff.
*Poleman 4883 (Smith Indic 34) ff. 9–11.
*Poleman 4924 (Smith Indic 52) 2 ff.
*Poleman 4946 (Smith Indic MB) XXIV f. 19.
*Poleman 4946 (Smith Indic MB) XXXV f. 1.
*Poleman 4946 (Smith Indic MB) XL ff. 15–22.
*Poleman 4946 (Smith Indic MB) LX ff. 1–2.
*Poleman 4946 (Smith Indic MB) LXX 2 ff.
*Poleman 4946 (Smith Indic MB) XCII ff. 2v. and 5.
*Poleman 4946 (Smith Indic MB) XCIII ff. 1–2.
*Poleman 4946 (Smith Indic MB) XCVI f. 1.
*Poleman 4946 (Smith Indic MB) CI f. 1.

II. Tables:

1. Sun. Table of true longitudes and daily motions for 0 to 365 days. The initial longitude at epoch is 2;9,8°; the daily motion for day 0 is 0;58,40°. The maximum daily motion is 1;1,23° for days 259 to 262; the minimum 0;56,54° for days 69 to 78. *Cf.* table 1 of the *Candrārkī*.

Manuscripts: 4904 (Smith Indic 100 ff. 3–7v); 4924 ff. 1–2v; 4946 XL ff. 15–17; 4946 XCII ff. 2v and 5.

2. Lord of the Year. Table of the lord of the year for 0 to 88 years. The parameter is 1;15,31,17,17 days per year (i.e., 1 year = 6,5;15,31,17,17 days); the epoch value is 6;0,39,30 days. To complete the cycle, add 1;24,38 days to the entry for k = 88. *Cf.* table 6 of the *Jagadbhūṣaṇa*.

Manuscripts: 4805 (Smith Indic 100 ff. 8–9); 4946 XL f. 15; 4946 XCIII f. 2.

3. Epact and Moon. Table of the epact and of the mean motion of the Moon for 0 to 121 years; the intercalary months are indicated opposite the appropriate years:

2	Āṣāḍha	32	Vaiśākha (var.: Jyeṣṭha)
5	Vaiśākha	34	Āśvina
7	Bhādrapada	37	Śrāvaṇa
10	Āṣāḍha	40	Jyeṣṭha
13	Jyeṣṭha	42	Āśvina
15	Āśvina	45	Śrāvaṇa
18	Śrāvaṇa	48	Āṣāḍha
21	Āṣāḍha	51	Jyeṣṭha
24	Vaiśākha	53	Āśvina
26	Bhādrapada	56	Āṣāḍha
29	Āṣāḍha	59	Jyeṣṭha

61 Āśvina
64 Śrāvaṇa
67 Āṣāḍha
70 Jyeṣṭha
72 Bhādrapada
75 Āṣāḍha
78 Jyeṣṭha
81 Vaiśākha
83 Śrāvaṇa
86 Āṣāḍha
89 Vaiśākha
91 Bhādrapada
94 Āṣāḍha
97 Jyeṣṭha
99 Āśvina
102 Śrāvaṇa
105 Āṣāḍha
108 Vaiśākha
110 Bhādrapada
113 Āṣāḍha
116 Jyeṣṭha
118 Āśvina
121 Śrāvaṇa

The parameter for the epact is 11;3,53,22,40 tithis a year; the epoch value is 6;26,49 tithis; and the bīja is − 0;5,28 tithis. The parameter for the Moon is 2,12;46, 40,32° per year; the epoch position is 1,17;21, 55°; and the bīja is − 1;5,35,24,38°. *Cf.* tables 7 and 8 of the *Jagadbhūṣaṇa.*

Manuscripts: 4916 (Smith Indic 100 ff. 9v–11v.); 4946 XL ff. 17v–18.

4. Moon and Lunar Anomaly. Table of mean motions for 0 to 60 ghaṭikās. The mean daily motions are: for the Moon, 13;10,35°, and for the lunar anomaly, 13;3, 54°. *Cf.* table 3 of the *Candrārkī,* and table 14 of the *Jagadbhūṣaṇa.*

Manuscripts: 4821 (Smith Indic 100 ff. 12–12v); 4946 XXIV f. 19; 4946 XL ff. 18v–19; 4946 XCIII f. 1.

5. Lunar Anomaly. Table of mean motion for 0 to 42 years. The parameter is 1,32;6,8,48° per year; the epoch position is 5,46;18,49°; and the bīja is − 0;24,38°. *Cf.* table 9 of the *Jagadbhūṣaṇa.*

Manuscripts: 4822 (Smith Indic 100 f. 13); 4946 XL f. 18v; 4946 XCIII f. 1v.

6. Moon and Lunar Anomaly. Table of longitudes for 0 to 365 days. Longitudes for day 0:

Moon	5,59;52,46°
Anomaly	5,59;52,51°.

Manuscripts: 4822 (Smith Indic 100 ff. 13v–20v); 4946 XL ff. 19–22.

7. Moon. Table of the lunar equation for arguments of 0° to 90° with differences between entries and differences between mean daily motions and true daily motions. Maximum equation is 5;2,10°; maximum increment or decrease in mean daily motion 1;8,54°. *Cf.* table 15 of the *Jagadbhūṣaṇa.*

Manuscripts: 4884 (Smith Indic 100 ff. 21–22v); 4946 XL ff. 22–22v.

The epoch of these tables is evidently the year 66 after the epoch of the *Jagadbhūṣaṇa* (Śaka 1560).

	text	bīja	Śaka 1626 according to the *Jagadbhūṣaṇa*
Lord of Year	6;0,39,30 days		6;0,39,31 days
Epact	6;26,49 tithis	− 0;5,28 tithis	6;28,24,42 tithis
Moon	1,17;21,55°	− 1;5,35,24,38°	1,17;40,56°
Anomaly	5,46;18,49°	− 0;24,38°	5,47;12,52°

This, then, corresponds to Saturday, 1 April 1704 in the Julian calendar.

8–12. Planetary tables.

These tables, like the planetary tables of the *Jagadbhūṣaṇa,* are based on the Babylonian Goal-year periods: 59 years for Saturn, 83 for Jupiter, 79 for Mars, 227 for Venus, and 46 for Mercury. There are five columns. Column 1(A) lists the 27 avadhis (1 to 27 = 0 to 26). Column 2(B) gives the true longitudes (2 per avadhi for Mercury). Column 3 (L) gives the differences between the entries in column 2 for cycle n and cycle n + 1; i.e., to find the longitudes for the year N = 59 from the epoch for Saturn, one adds together the entries in columns 2 and 3 for N = 0. Column 4 (D) gives the progress of the planet on the first day of the avadhi. And column 5 (M) gives the differences between the entries in column 4 for cycle n and cycle n + 1. G is in the margin, along with H, which gives the differences between entries in G for cycle n and cycle n + 1.

We can compute the longitude of each planet for the beginning of the first year of cycle 2 by multiplying the entry in column 4 for k = 27 in the last year of cycle 1 by 1;15,30 (the approximate difference between a sidereal year and 364 days), and adding the product to the entry in column 2 for k = 27 in that year. The difference between the result of this computation and the entry in column 2 for k = 1 in the year 0 of cycle 1 should be equal to the entry in column 3 for k = 1 in the year 0 of cycle 1.

Saturn:	N = 58, k = 27	col. 2 =	0^{s}3;31,34°
		col. 4 =	0;6,37°
		computation =	0^{s}3;39,54°
	N = 0, k = 1	col. 2 =	0^{s}3;4,26°
		Δ =	0;35,28°
	N = 0, k = 1	col. 3 =	0;37,42°
Jupiter:	N = 82, k = 27	col. 2 =	1^{s}3;47,2°
		col. 4 =	− 0;13,52°
		computation =	1^{s}3;29,35°
	N = 0, k = 1	col. 2 =	1^{s}4;7,41°
		Δ =	− 0;38,6°
	N = 0, k = 1	col. 3 =	− 0;44,14°
Mars:	N = 78, k = 27	col. 2 =	9^{s}13;17,47°
		col. 4 =	0;40,46°
		computation =	9^{s}14;9,5°
	N = 0, k = 1	col. 2 =	9^{s}13;21,46°
		Δ =	0;47,19°
	N = 0, k = 1	col. 3 =	0;47,30°
Venus:	N = 226, k = 27	col. 2 =	10^{s}23;51,47°
		col. 4 =	1;12,22°
		computation =	10^{s}25;22,51°

	N = 0, k = 1	col. 2 = 10ˢ25;22,54°
		Δ = −0;0,3°
	N = 0, k = 1	col. 3 = +0;0,13°
Mercury:	N = 45, k = 27	col. 2 = 11ˢ11;17,20°
		col. 4 = 1;44,33°
		computation = 11ˢ13;28,54°
	N = 0, k = 1	col. 2 = 11ˢ13;47,30°
		Δ = −0;18,36°
	N = 0, k = 1	col. 3 = −0;18,14°

Now we can compute the mean yearly motion, by adding column 3 for N = 0, k = 1 to the revolutions of the planet in a cycle, and dividing by the years in a cycle.

Saturn	12;12,50,32°
Jupiter	30;21,9,13°
Mars	3,11;24,8,43°
Venus (anomaly)	3,48;22°
Mercury (anomaly)	54;46°

At the epoch—Saturday, 1 April 1704 in the Julian calendar—the positions of the planets as compared with positions computed by modern tables were:

	Sidereal	Tropical	Sidereal-Tropical
Saturn	3;4,26°	25°	−22°
Jupiter	34;7,41	54	−20
Mars	4,43;21,46	302	−19
Sun	2;9,8	22	−20
Venus	5,25;22,54	346	−21
Mercury	5,43;47,30	0	−16
Moon	1,16;16,20		

Manuscripts:

8. Mars: 4766 ff. 1–20v (N = 0 − 78)
9. Mercury: 4766 ff. 21–44v (N = 0 − 45); 4946 XXXV f. 1 (N = 30).
10. Jupiter: 4766 ff. 45–65v (N = 0–82).
11. Venus: 4766 ff. 66–122v (N = 0–226).
12. Saturn: 4766 ff. 123–137v (N = 0–58).

III. Tables for specific years based on the Anonymous of 1704:

Saṃ. 1849/Śaka 1714 (A.D. 1792): 4946 LX ff. 1–2.
Saṃ. 1852/Śaka 1717 (A.D. 1795): 4946 XCVI f. 1.
Saṃ. 1854/Śaka 1719 (A.D. 1797): 4946 LXX 2 ff.
Saṃ. 1865/Śaka 1730 (A.D. 1808): 4865 f. 1.
Saṃ. 1866/Śaka 1731 (A.D. 1809): 4865 f. 2.
Śaka 1769 (A.D. 1847): 4946 CI f. 1.

ANONYMOUS OF 1714

I. Manuscripts:

*Poleman 4878 (Smith Indic 21) 2 ff.

II. Tables:

1. Table of yearly mean motions of the Epact, the Lord of the Year, Mars, Mercury, Jupiter, Venus, Saturn, Lunar Node, Moon, and Lunar Anomaly from Śaka 1636 to 1697 (A.D. 1714–1775); the entries are complete through Śaka 1693 (A.D. 1771). For Śaka 1694–1697 (A.D. 1772–1775) there are entries for Epact, Lord of the Year, Moon, and Lunar Anomaly; for Śaka 1694 an entry for Lunar Node as well.

The entries are very close to those computed from the *Mahādevī* for the same years; here are given only the entries for Śaka 1636 (A.D. 1714).

	text	computed from *Mahādevī*
Epact	27;7,17,35 tithis	27;7,18,32 tithis
Lord of Year	4;35,52,25 days	4;35,52,23,2 days
Mars	0;38,51,16(×6°)	0;38,51,31,6(×6°)
Mercury	22;37,56,29	22;37,56,50,38
Jupiter	56;42,32,4	56;42,32,4,22
Venus	0;39,15,32	0;39,15,34,36
Saturn	22;0,28,50	22;0,29,57,12
Lunar Node	7;20,29,36	7;21,33,38,39,48
Moon	10;25,27,10	[10;25,27,42,32]
Lunar Anomaly	6;7,37,17	

The epoch date is Thursday, 1 April 1714 in the Julian calendar.

	Sidereal	Tropical	Sidereal Tropical
Saturn	2,12;2,53,0°	155°	−23°
Jupiter	5,40;15,12,24	3	−23
Mars	3;53,7,36	29	−25
Venus' anomaly	3;55,33,12	(Venus) 22	
Mercury's anomaly	2,15;47,38,54	(Mercury) 39	
Moon	1,2;32,43,0		
Lunar Anomaly	36;45,43,42		
Lunar Node	5,15;57,2,24	c.255	

JAYAVINODASĀRAṆĪ

These tables were written for the famous Rajput Mahārāja of Amber, Jayasiṃha (1693–1743), who built the observatories of Delhi, Jaipur, Ujjain, Benares, and Mathurā (G. R. Kaye, *The Astronomical Observatories of Jai Singh, ASI*, Imp. Ser. 40, Calcutta 1918); his *Yantrarājaracanā* (otherwise known as the *Yantrarājakārikā*) has been published by Kedāranātha in a journal published in Jayapura called *The Pandit* (1, 1924) with a note by A. ff. Garrett, and again by (another?) Kedāranātha with his own commentary as *Rājasthāna Purātana Granthamālā* 5, Jayapura 1953. Also composed for Jayasiṃha were the *Rekhāgaṇita* based on Naṣīr al-Dīn al-Ṭūsī's recension of the Arabic translation of Euclid's Στοιχεῖα (ed. H. H. Dhruva and

K. P. Trivedī, *Bombay SS* 61–62, Bombay 1901–1902; *cf.* also L. J. Rocher, "Euclid's Stoicheia and Jagannātha's Rekhāgaṇita", *JOIBaroda* 3,1953/4, 236–256) and the *Samrāṭsiddhānta,* a Sanskrit version of Naṣīr al-Dīn's version of Ptolemy's Σύνταξις Μαθηματική; both translations were made by Jagannātha, that of the *Rekhāgaṇita* in 1726 and that of the *Samrāṭsiddhānta* in 1732. Jayasiṃha was also responsible for the Persian work, *Zīj i jadīd i Muḥammad-Shāhī,* which was completed in 1728; his "advisor" in composing this work was Abū al-Khayr Khayr Allāh Khān (C. A. Storey, *Persian Literature,* vol. 2, pt. 1, London 1958, pp. 93–95). The present work was apparently written in 1735.

I. Manuscripts:

Calcutta Sanskrit College 17 19 ff.
*Poleman 5107 (Harvard 61) 23 ff.

II. Tables:

1. Table of yogas for Śaka 1657, 1687, 1717, 1747, 1777, and 1807 (A.D. 1735, 1765, 1795, 1825, 1855, and 1885). Four functions are tabulated: the yogadhruva, the week-day on which the yoga begins (yogavārādi), the position in the solar cycle of 387;57,30,4 days (yoge ravikendra), and the position in the lunar cycle of 29;16,1,10 days (yoge candrakendra).

Manuscripts: 5107 f. 1v.

2. Table of yogas for 1 to 30 years. The yearly motions are:

yogadhruvas:	10	modulo 27
yogavāras:	1;17,52,42	modulo 7
yoge ravikendra:	0;2,29,56	modulo 7
yoge candrakendra:	7;31,43,8	modulo 29;16,1,10.

Manuscripts: 5107 ff. 2–2v.

3. Table of yogas for 0 to 388 days; this gives only the last two functions. At 388 days they are, respectively, 1;18,18 and 7;31,46.

Manuscripts: 5107 ff. 3–6.

4. Equations of the candrakendra for 0 to 29 days horizontal, 0 to 59 ghaṭikās vertical. The maximum equation is 21,22.

Manuscripts: 5107 ff. 7–8v.

5. Equations of the ravikendra for 0 to 388 days horizontal, 0 to [59] ghaṭikās vertical. The maximum equation is 10,1.

Manuscripts: 5107 ff. 9–23v.

ANONYMOUS OF 1741

I. Manuscripts:

*Poleman 5177 (Smith Indic 28) ff. 8–23. Saṃ. 1846 = A.D. 1789.
*Poleman 4770 (Smith Indic 179) ff. 4–16.
*Poleman 4946 (Smith Indic MB) XLVI f. 6.
*Poleman 4946 (Smith Indic MB) C f. 1.

II. Tables:

1. Table of the Epact for 30 periods of 30 years, and for 0 to 29 years. The parameter is 11;3,53,22,40 tithis per year; the epoch position 26;53,35,5,2 tithis.

Manuscripts: 4770 f. 4; 5177 f. 8.

2. Table of the Lord of the Year set up as is table 1. The parameter is 1;15,31,17,17 days per year; the epoch position 3;34,58,52,48 days. These epoch positions indicate a date of Wednesday, 1 April 1741 in the Julian calendar.

Manuscripts: 4770 f. 4; 5177 f. 8v.

3. Table of the tithidhruva (modulo 30) for 30 periods of 30 years, and for 0 to 29 years. The parameter is 10;56,6,37,20 tithis per year; the epoch position 26;6,24,54,58 tithis. The parameter is equal to 11 – 0;3,53,22,40 tithis.

Manuscripts: 4770 f. 4v; 5177 f. 9.

4. Table of the nakṣatrayogadhruva (modulo 27) for 30 periods of 30 years, and for 0 to 29 years. The parameter is 10;2,29,57,36 nakṣatras per year; the epoch position 24;41,46,25,28 nakṣatras. The parameter is derived by subtracting from 11 the sexagesimal fraction resulting from the division of the Moon's yearly motion by the length of a nakṣatra.

$$2,12;46,40,30 \div 13;20 = 9;57,30,2,24$$
$$11 - 0;57,30,2,24 = 10;2,29,57,36.$$

Manuscripts: 4770 f. 4v; 5177 f. 9v.

5. Table of the tithikendra (modulo 27;59,33) for 30 periods of 30 years, and for 0 to 29 years. The parameter is 7;5,48,29,45 tithis per year; the epoch position 12;10,45,40,52 tithis. The parameter indicates that 13 anomalistic months equal 6,11 – 7;5,48,29,45 tithis, or 6,3;54,11,30,15 tithis. One anomalistic month, then, equals 27;59,33,11,33 tithis.

Manuscripts: 4770 f. 5; 5177 f. 10.

6. Table of the nakṣatrakendra (modulo 27;13,48) for 30 periods of 30 years, and for 0 to 29 years. The parameter is 7;0,30,12,0 nakṣatras per year; the epoch position 12;30,33,24,57 nakṣatras. The parameter indicates that 13 anomalistic months equal 6,1 – 7;0,30,12 nakṣatras, or 5,53;59,29,48 nakṣatras. One anomalistic month, then, equals 27;13,48,26,46 nakṣatras.

Manuscripts: 4770 f. 5; 5177 f. 10v.

7. Table of the yogakendra (modulo 29;16,1) for 30 periods of 30 years, and for 0 to 29 years. The parameter is 7;31,45,53,49 yogas per year; the epoch position 13;20,47,36,47 yogas. The parameter indicates that 13 anomalistic months equal 6,28 – 7;31,45,53,49 yogas, or 6,20;28,14,6,11 yogas. One anomalistic month, then, equals 29;16,1,5,5 yogas.

Manuscripts: 4770 f. 5v; 5177 f. 11.

We can now compare these anomalistic months:
1 tithi = 0;59,3,40,11 days; therefore, 1 anom. mo. = 27;33,16,21,51 days.

1 nakṣatra = 1;0,42,53,20 days; therefore, 1 anom. mo. = 27;33,16,18,57 days.

1 yoga = 0;56,29,21,42 days; therefore, 1 anom. mo. = 27;33,16,20,21 days.

8. Table of the tithivāra (modulo 7) for 30 periods of 30 years, and for 0 to 29 years. The parameter is 1;11,41,33,24,30 days per year; the epoch position 3;18,45,46,55 days. The parameter indicates that 1 tithi equals 0;59,3,40,11,52 days.

Manuscripts: 4770 f. 5v; 5177 f. 11v.

9. Table of the nakṣatravāras (modulo 7) for 30 periods of 30 years, and for 0 to 29 years. The parameter is 1;18,3,1 days; the epoch position 3;56,30,8,34 days. The parameter indicates that 1 nakṣatra equals 1;0,42,53,20 days.

Manuscripts: 4770 f. 6; 5177 f. 12.

10. Table of the yogavāras (modulo 7) for 30 periods of 30 years, and for 0 to 29 years. The parameter is 1;17,52,25 days; the epoch position 3;55,1,52,4 days. The parameter indicates that 1 yoga equals 0;56,29,21, 42 days.

Manuscripts: 4770 f. 6; 5177 f. 12v.

11. Table of the tithikendrakṣepakas (modulo 27;59,33) and of the tithivāras (modulo 7) for 0 to 37 periods of 10 days each. Also given are columns of cālakas, which are the differences between successive entries measured in units equal to 1/6 of a ghaṭikā, or 10 palas. This is table 5 of the *Pañcāṅgavidyādharī.*

Manuscripts: 4770 f. 6v; 5177 f. 13.

12. Table of the nakṣatrakendrakṣepakas (modulo 27; 13,48) and of the nakṣatravāras (modulo 7) for 0 to 41 periods of 9 days each. Also given are columns of cālakas measured in units of 10 palas. This is table 6 of the *Pañcāṅgavidyādharī.*

Manuscripts: 4770 f. 7; 5177 f. 13v.

13. Table of the yogakendrakṣepakas (modulo 29;16,1) and of the yogavāras (modulo 7) for 0 to 43 periods of 9 days each. Also given are columns of cālakas measured in units of 10 palas. This is table 7 of the *Pañcāṅgavidyādharī.*

Manuscripts: 4770 f. 7v; 5177 f. 14.

14. Table of week-days on which the Sun enters the 27 nakṣatras.

Aśvinī	363	4;47,51
Bharaṇī	11	4;28,59
Kṛttikā	25	4;16,51
Rohiṇī	39	4;10,48
Mṛgaśiras	53	4;9,17
Ārdrā	67	4;11,12
Punarvasu	81	4;14,34
Puṣya	95	4;17,34
Āśleṣā	109	4;18,18
Maghā	123	4;15,53
Pūrvaphālgunī	137	4;6,54
Uttaraphālgunī	150	3;52,3
Hasta	164	3;30,21
Citra	178	3;1,24
Svāti	191	2;25,25
Viśākhā	204	1;42,5
Anurādhā	217	0;54,56
Jyeṣṭhā	231	0;2,15
Mūla	244	6;6,19
Pūrvāṣāḍhā	257	5;6,40
Uttarāṣāḍhā	270	4;10,35
Śravaṇa	283	3;13,31
Dhaniṣṭhā	296	2;19,3
Śatabhiṣak	309	1;28,17
Pūrvabhādrapadā	322	0;42,26
Uttarabhādrapadā	336	0;2,38
Revatī	344[349]	6;22,29

Manuscripts: 5177 f. 14v; 4946 XLVI f. 6; 4946 C f. 1.

15. Table of week-days on which the Sun enters the 12 zodiacal signs, and of the lengths of daylight on those days.

Aries	363	4;47,51	31;8
Taurus	28	0;44,44	32;34
Gemini	60	4;9,50	33;22
Cancer	91	0;46,52	33;22 [33;10]
Leo	123	4;15,13	32;6
Virgo	154	0;17,14	30;34
Libra	184	2;44,16	28;54 [28;52]
Scorpio	214	4;37,26	27;26
Sagittarius	244	6;6,19	26;38
Capricorn	273	0;26,9	26;50
Aquarius	302	1;53,7	27;54
Pisces	332	3;42,3	29;28 [29;26]

Manuscripts: 5177 f. 14v; 4946 XLVI f. 6v.

16. Table of corrections to the tithikendra, normed so as to be always positive. The horizontal argument is 0 to 27 days of an anomalistic month; the vertical argument 0 to 59 ghaṭikās. The maximum entry is 49;34, the minimum 0;0, and the mean 24;47.

Manuscripts: 4770 ff. 8–10v; 5177 ff. 15–17v.

17. Table of corrections to the nakṣatrakendra set up as is table 16. The maximum entry is 45;52, the minimum 0;0, and the mean 22;56.

Manuscripts: 4770 ff. 11–13v; 5177 ff. 18–20v.

18. Table of corrections to the yogakendra set up as is table 16 save that the horizontal argument is 0 to 29 days of a synodic month. The maximum entry is 42;40, the minimum 0;0 and the mean 21;20.

Manuscripts: 4770 ff. 14–16v; 5177 ff. 21–23v.

ANONYMOUS OF 1747

I. Manuscripts:

*Poleman 4946 (Smith Indic MB) XX 1 f. Saṃ. 1835, Śaka 1700 = A.D. 1778.

II. Tables:

1. Tables of mean motions of the Lord of the Year, Epact, Moon, and Lunar Anomaly for Saṃ. 1804 to 1865 (A.D. 1747 to 1808). The parameters and epoch positions are:

	Yearly motion	Epoch position
Lord of the Year	1;15,31,17,20 days	4;11,24,54,50 days
Epact	11;3,53,22,40 tithis	2;18,53,20,50 tithis
Moon	2,12;46,40,32°	1;20,42,9°
Lunar Anomaly	1,32;6,8,48	5,45;45,44,33

The epoch, then, is Thursday, 2 April 1747 in the Julian Calendar.

Manuscripts: 4946 XX ff. 1–1v.

GRAHALĀGHAVASĀRIṆĪ

I. Manuscripts:

*Poleman 4694 (Smith Indic 26) 6 ff.

II. Tables:

With this work based on Gaṇeśa's *Grahalāghava* compare the *Grahalāghavīyamadhyamaspaṣṭārkasāriṇī*. The epoch of the present tables is Saturday, 2 April 1754 in the Julian calendar.

	Sidereal	Tropical	Sidereal-Tropical
Lord of the Year	6;0,46 days		
Epact	19 tithis		
Saturn	4,23;56,57°	287°	-23°
Jupiter	1,55;7,15	128	-13
Mars	1,40;57,27	92	+ 9(!)
Venus' anomaly	11;30,15	(Venus) 27	
Mercury's anomaly	2,35;51,11	(Mercury) 29	
Moon	3,50;3,46		
Lunar Anomaly	2,23;21,26		
Lunar Node	3,24;36,40	c.200	

A. The tables of mean motions are for the following units of time:

10-year periods from Śaka 1676 to 1776 (A.D. 1754 to 1854). Lord of the Year, Precession, Epact (in integer tithis), Moon, Lunar Anomaly, Lunar Node, Mars, Mercury's anomaly, Jupiter, Venus' anomaly, and Saturn. The parameter for precession is 1° every 60 years; the epoch amount—20;31°—indicates that precession is assumed to have been 0° in Śaka 444 (A.D. 522). This date is also used in the *Bṛhatpārāśarahorāśāstra*, Pūrvakhaṇḍa 3,31.

From 1 to 10 years. Lord of the Year (1;15,31,30 days per year), Epact (in integer tithis) (11 tithis per year), Moon, Lunar Anomaly, Lunar Node, Mars, Mercury's anomaly, Jupiter, Venus' anomaly, and Saturn.

From 100 to 300 days and from 10 to 90 days. Sun, Moon, Lunar Anomaly, Mars, Mercury's anomaly, Jupiter, Venus' anomaly, Saturn, and Lunar Node.

From 1 to 9 days. Sun, Moon, Lunar Anomaly, Mars, Mercury's anomaly, Jupiter, Venus' anomaly, Saturn, and Lunar Node. The approximate mean daily motions are:

Saturn	0;2,0,24°
Jupiter	0;4,59,6
Mars	0;31,26,30
Sun	0;59,8,6
Venus (anom)	0;36,59,54
Mercury (anom)	3;6,24,6
Moon	13;10,34,48
Lunar Anomaly	13;3,54
Lunar Node	-0;3,10,48

Manuscripts: 4694 ff. 1–3.

B. Tables of the equation of the center for the Sun and Moon. There are 6 columns (5 and 6 are omitted from the lunar table) in these tables. (1). The argument from 0° to 90° in intervals of 5°. (2). The equation given to minutes and seconds. (3). The difference for each degree of increase in the argument, computed by dividing the differences between entries in column 2 by 5. (4). The increment to (or decrease from) the mean daily motion given to seconds. (5). The difference for each degree of increase in the argument. (6). The bhujaphala of the Moon; not clear. The maximum equations of the Sun and Moon are respectively 2;14,45° and 5;1,40°; the solar apogee is placed at Gemini 18°.

Manuscripts: 4694 ff. 3–3v.

C. Tables of the equations of the center for the other planets. In this table the argument increases in intervals of 15° from 0° to 75°; three functions are tabulated for each planet, corresponding to columns 2,3, and 4 in the solar and lunar tables. The maximum equations and apsidal longitudes are:

	Emax	Apogee
Saturn	9;18°	240°
Jupiter	5;42	180
Mars	12;54	120
Venus	1;30	90
Mercury	3;36	210

Manuscripts: 4694 ff. 3v–4.

D. Tables of the equations of the conjunctions of the five star-planets. The argument again increases in intervals of 15°, but from 0° to 165°; and an extra table is added giving the equations of Mars, Mercury, and Venus for single degrees from 166° to 189°. The same three functions are tabulated for each planet as is the case in set C. The maximum equations and the longitudes at which they occur are:

	Emax	Longitudes
Saturn	5;42°	90° to 105°
Jupiter	10;48	105
Mars	40	135
Venus	46;6	135
Mercury	21;12	105 to 120.

Manuscripts: 4694 ff. 4–4v.

E. Tables of oblique ascensions of the signs measured in ghaṭikās, of the interval measured in signs and degrees between the first degree of each sign when it is in the ascendent and the degree which at that time is in mid-heaven, and of the rising-times of the signs measured in half-palas. The latitudes for which these tables are computed are ⟨Laṅkā⟩ (the equator), Gūrjaradeśa (Gujarat) Pāṭaṇapura (Pattan), Sthambhatīrtha (Cambay), and Amadāvāda (Ahmadabad). A last table (which omits the second function) substitutes for the first the amount of time since the rising of Aries 0° that has elapsed when the first degree of each zodiacal sign is in mid-heaven.

Manuscripts: 4694 f. 5.

ANONYMOUS OF 1790

I. Manuscripts:

*Poleman 4946 (Smith Indic MB) XI 1 f.

II. Tables:

1. Table of the yogadhruva and the yogakendra for periods of 24 years from Śaka 1712 to 1808 (A.D. 1790 to 1886). There are 3 columns after the column of arguments: the yogas, the week-days, and the yogakendras. The respective parameters are: 23 yogas (modulo 27); 2;12,29 days (modulo 7); and 4;6,15 (modulo 30). The epoch positions for Śaka 1712 (A.D. 1790) are:

yogadhruva	26 yogas
week-day	2;58,13 days
yogakendra	29;50,⟨59⟩

Manuscripts: 4946 XI f. 1.

2. Table of the yogadhruva for 1 to 24 years. Again there are the same 3 columns after the column of arguments. The yearly parameters are:

yogadhruva	9;57,30 yogas (modulo 29)
week-day	1;15,31 days
yogakendra	7;29,16 (modulo 29;16,1)

Manuscripts: 4946 XI f. 1.

3. Table of week-days with which the first tithis of 1 to 13 synodic months begin and their corresponding cālakas, and of the tithikendras and their corresponding cālakas for 1 to 13 synodic months. With the first part compare table 3 of the Anonymous of 1638; the second part of the table is identical with table 1 of the Anonymous of 1638. The lengths of the months vary. The tithikendras are modulo 27;59,33. The week-day for month 1 is 6;10,28.

Manuscripts: 4946 XI f. 1.

4. Table of week-days with which the first nakṣatras of 1 to 14 sidereal months begin and their corresponding cālakas, and of the nakṣatrakendras and their corresponding cālakas for 1 to 14 sidereal months. With the first part compare table 4 of the Anonymous of 1638; the second part is identical with table 2 of the Anonymous of 1638. The lengths of the months vary. The nakṣatrakendras are modulo 27;13,48. The week-day for month 1 is 6;0,13.

Manuscripts: 4946 XI f. 1.

5. Table of week-days with which the first yogas of 1 to 15 periods begin and their corresponding cālakas, and of the yogakendras and their corresponding cālakas for 1 to 15 periods. With the first part compare table 7 of the Anonymous of 1638; the second part is identical with table 6 of the Anonymous of 1638. The lengths of the months vary. The yogakendras are modulo 29;16,1. The week-day for month 1 is 5;50,58.

Manuscripts: 4946 XI f. 1v.

KARAṆAKESARI

There are several works entitled *Karaṇakesari* in Sanskrit:

1. Attributed to Bhâskara. Manuscripts: Baroda 11268 2 ff. Saṃ. 1819 = A.D. 1762; CP, Hiralal 677 (from Kârañjâ in Akolâ District); and PL, Buhler 17 13 ff.

2. Attributed to Râma. Manuscripts: PL, Buhler 18 Saṃ. 1820 = A.D. 1763; and N-W Prov X (1886) 30 2 ff. (from Azamgarh).

3. Written by Īśvara at Kollamburapura during the reign of Firûz Shâh (1351–1388). Manuscripts: Anup 4456 26 ff. Śaka 1465 = A.D. 1543.

4. An anonymous text. Manuscripts: Goṇḍal 14 17 ff. Saṃ. 1704 = A.D. 1647.

To which, if to any, of the above the *Karaṇakesari* which is under consideration here may be related is unknown.

I. Manuscripts:

*Poleman 4946 (Smith Indic MB) XIV ff. 3–11.
*Poleman 4946 (Smith Indic MB) XXVII f. 1.

II. Tables:

1. Table of the elongation between the mean Sun and the Lunar Node for 1 to 130 periods of 130 years. In 130 years, the mean Sun diminished by the Lunar Node equals 5,56;43,40°; and in 130 × 130 years it equals 4,54;36,40°.

Manuscripts: 4946 XIV f. 3.

2. Table of the elongation between the mean Sun and

the Lunar Node for 0 to 130 years. The yearly parameter is 19;21,34°; the epoch position 3,29;24,36°.

Manuscripts: 4946 XIV f. 3v.

3. Table of the elongation between the mean Sun and the Lunar Node for 1 to 27 avadhis. There are 4 columns. Column 1 lists the avadhis, column 2 indicates the elongation, column 3 the daily progress of the function on the first day of the avadhi, and column 4 the cālakas. The initial position is 2;19,55°.

Manuscripts: 4946 XIV f. 4.

4. Table of the "digits of the arrow" (? śarāṅgula); the argument is the bhujāṃśa or anomaly in degrees. There are three columns in the table. The first gives the argument from 1° to 16°; the second gives the śarāṅgulas, which increase from 0;10 to 24;45 digits; and the third gives the differences between the entries in column 2. A note in the margin remarks that with the "sphere of the arrow" (? śaragola) one discovers the direction: north in Aries through Virgo, south in Libra through Pisces. This apparently has something to do with the projection of eclipses; see tables 13 and following.

Manuscripts: 4946 XIV f. 4.

5. Table of the apparent solar diameter (ravibimba) measured in digits (aṅgulas) for various daily elongations of the Sun and the Lunar Node ranging from 0;59,56° to 1;4,42° at intervals of varying amounts; there are 20 entries in all. The minimum apparent diameter (at 0;59,56° or a daily progress of the Sun of 0;56,45°) is 10;12 digits; the maximum (at 1;4,42°, or a daily solar progress of 1;1,31°) 11;11 digits.

Manuscripts: 4946 XIV f. 4.

6. Table of the apparent lunar diameter and the diameter of the cone of the earth's shadow measured in digits; these diameters depend on the Moon's distance, which is assumed to vary inversely with its velocity. The argument with which one enters this table is the length of the current tithi, which ranges from 0;52 days to 1;7 days. The maximum diameters of the Moon and the earth's shadow are respectively 11;57 and 30;45 digits when a tithi equals 0;52 days; the minimum respectively 9;30 and 23;11 digits when a tithi equals 1;7 days.

Manuscripts: 4946 XIV f. 4v.

7. Table of corrections to be added algebraically to the diameter of the cone of the earth's shadow due to the varying distance of the Sun, which depends on its position in the zodiacal relative to its apogee. The table is as follows:

+		−
Aries	0;11	Libra
Taurus	0;16	Scorpio
Gemini	0;20	Sagittarius
Cancer	0;16	Capricorn
Leo	0;11	Aquarius
Virgo	0;0	Pisces

Manuscripts: 4946 XIV f. 4v.

8. Table of the half-duration of a lunar eclipse expressed in ghaṭikās; the argument is the diameter of the eclipsed body expressed in digits from 1 to 21. Thus, the entry for 1 digit is 0;57,30 ghaṭikās; the entry for 21 digits is 55,17, which means that the half-duration of the eclipse is 1;55,17 ghaṭikās.

Manuscripts: 4946 XIV f. 4v.

9. Table of the duration of totality of a lunar eclipse measured in ghaṭikās; the argument is 1 to 9 digits beyond the minimum required for totality to occur. The entries seem to be corrupt.

Manuscripts: 4946 XIV f. 4v.

10. Table of true longitudes and daily progresses of the Sun on the first day of 1 to 27 avadhis. The initial solar longitude is Aries 2;9,50°.

Manuscripts: 4946 XIV f. 4v.

11. Table of differences between the entries in table 8 for 1 to 21 digits.

Manuscripts: 4946 XIV f. 4v.

12. Table of "thirds of duration of the eclipse" (? sthityagnibhāga) for 1 to 21 digits of the eclipsed body of the Moon. The entries increase from 0,31 to 1,35.

Manuscripts: 4946 XIV f. 5.

13. Table of "correction to the arrow" (? bāṇaphala and śaraphala); the argument is called the "digits of the mean arrow" or "digits of the arrow of the meridian" (madhyabāṇāṅgula) from 1 to 21. The entries increase from 0,1 to 0,27. *Cf.* the śarāṅgulas in table 4.

Manuscripts: 4946 XIV f. 5.

14. Table of deflection (valana) measured in degrees; the argument is the hour-angle measured in degrees (natāṃśa). The table has three columns. The first gives the argument for 0° to 45°; the second the deflection, which decreases from 22;35,39° to 0;0,3°; and the third the differences between entries in column 2.

Manuscripts: 4946 XIV ff. 5–5v.

15. Table of the lords of the syzygies (parveśa); the argument is called "signs of the Node, the Moon, and the Sun" (saṃpātacandrasūryarāśi). A complete copy of the table is as follows:

0	Varuṇa	11	Yama	12 [22]	Brahmā
5	Śaśi	17	Varuṇa	23	Indra
6	Indra	18	Agni	24	Kubera

Manuscripts: 4946 XIV f. 5v.

16. Table of deflection (valana) expressed in degrees at the time of first contact or that of last contact; the argument is the koṭyaṃśa or 90°—(anomaly in degrees), with the tropical longitude of the eclipsed body being used. See table 4 which employs the bhujāṃśa.

There are three columns: the first lists the argument from 0° to 90°; the second the deflection, which increases from 0;0,3° to 24;0,1°; and the third the differences between entries in column 2. It is further stated that, if the tropical longitude of the eclipsed body is in the semicircle from Cancer through Sagittarius, the deflection is to the South; in that from Capricorn through Gemini, to the North.

Manuscripts: 4946 XIV f. 6.

17. Table of the "corrected deflection" (valana spaṣṭa) of the Sun. The argument is from 0 to 47 (degrees?), and the entries increase from 0;5 to 8;2 (digits?).

Manuscripts: 4946 XIV f. 6v.

18. Table of the "corrected deflection" of the Moon. Again the argument is from 0 to 47 (degrees?); the entries increase from 0;4 to 13;51 (digits?).

Manuscripts: 4946 XIV f. 6v.

19. Table of half-lengths of daylight measured in ghaṭikās for the Sun's being in 0° to 29° horizontal of 0 to 11 zodiacal signs vertical. The tropical longitude of the Sun is referred to. The longest day is 33;34 ghaṭikās at Cancer 0°.

Manuscripts: 4946 XIV f. 7.

20. Table of accumulated rising-times measured in ghaṭikās of 0° to 29° horizontal of 0 to 11 zodiacal signs vertical; longitudes are again tropical.

Manuscripts: 4946 XIV ff. 7v–8.

21. A table based on the following division of a day (120 parts):

Aries Pisces	7;36
Taurus Aquarius	8;38
Gemini Capricorn	10;12
Cancer Sagittarius	11;20
Leo Scorpio	11;18
Virgo Libra	10;56

The table is for 1 to 60 equal parts of these numbers horizontal, and for 1 to 6 pairs of zodiacal signs vertical.

Manuscripts: 4946 XIV ff. 8v–9v; 4946 XXVII ff. 1–1v.

22. Table of parallax of longitude measured in ghaṭikās; the argument (for 0° to 91°) is the angular difference between the Moon and the horizon-point rather than that between the Moon and the highest point of the ecliptic. Parallax for an argument of 0° is 3;40 ghaṭikās (1 hour 28 minutes); that for 24° (= 90° − 66°) is the maximum, 4;0 ghaṭikās (1 hour 36 minutes); and that for 89° is 0;7 ghaṭikās (2 minutes 48 seconds). *Cf.* tables 77–77a of al-Khwārizmī.

Manuscripts: 4946 XIV f. 9v.

23. Table of the "multiplier for correcting the parallax" (lambanaspaṣṭaguṇaka); the argument is the tropical longitude of the ascendent. The table is arranged for 0° to 29° horizontal of 0 to 11 zodiacal signs. The maximum "multiplier" is 35,59 at Virgo 9°–12° and Libra 16°–21°; the minimum 27,21 at Pisces 29°.

Manuscripts: 4946 XIV f. 10.

24. Table of the parallax in latitude (nati) measured in digits; the argument is the tropical longitude. The table is set up for 0° to 29° of 0 to 11 zodiacal signs. The maximum entry is 11;46 digits around Aries 0°, the minimum 0;1 digits at Virgo 12° and Libra 18°.

Manuscripts: 4946 XIV f. 10v.

25. Table of the half-duration of a solar eclipse expressed in ghaṭikās; the argument is the number of digits from 1 to 12 of the diameter of the solar disc which are eclipsed at mid-eclipse. The values increase from 1;7 ghaṭikās for 1 digit to 2;42 ghaṭikās for 11 digits; the entry for 12 digits is mistakenly given as 0;0.

Manuscripts: 4946 XIV f. 11.

It is doubtful that the next two tables come from the *Karaṇakesarī.*

26. Table entitled: "Table of the vikalās of the planets in horoscopy; measured in days." It is a multiplication table for 0;1,48 multiplied by 1,2,3,. . . .60; the text has that the product of 0;1,48 × 60 is 0;0,0.

Manuscripts: 4946 XIV f. 11.

27. Table of attributes of the 28 nakṣatras. For each nakṣatra is recorded its animal (horse, elephant, goat, snake, etc.), its "world" (that of the gods, of men, or of demons), and its altitude (low, middling, or high).

Manuscripts: 4946 XIV f. 11v.

BHĀVAPATTRA

I. Manuscripts:

*Poleman 4903 (Smith Indic 96) 2 ff.

II. Tables:

The argument of this table is each degree (0° to 29°) of each sign of the zodiac, apparently at midheaven (the cusp of the 10th place); the entries in column 2, then, correspond to the ascendent point. This part of the system, therefore, is similar to that of Theon's *Handy Tables* and of the *Zīj al-Sindhind* of al-Khwārizmī (ed. A. Bjørnbo, R. Besthorn, and H. Suter, København 1914, ch. 25 on pp. 18–19 and table 59–59b on pp. 171–173; see the translation and commentary by O. Neugebauer, *The Astronomical Tables of Al-Khwārizmī,* København 1962, pp. 46–48 and 104–105). It is said that the entry in column 3 corresponds to the hypogee or cusp of the 4th place; this, however, should be exactly 180° from the argument (the cusp of the 10th place), but is not.

DINAMĀNAPATTRĀṆI

These tables indicate for every degree of solar motion the corresponding length of daylight, usually in ghaṭikās.

They are usually set up with the 12 zodiacal signs as vertical arguments, and the 30° (0° to 29°) of a sign as horizontal argument. For their computation, see O. H. Schmidt, "The Computation of the Length of Daylight in Hindu Astronomy," *Isis* 35, 1944, 205–211, who brilliantly examined 4894.

Manuscripts:

*Poleman 4858 (Smith Indic 95) 1 f.
*Poleman 4889 (Smith Indic 113) 2 ff.
*Poleman 4894 (Smith Indic 23) A 1 f.
*Poleman 4894 (Smith Indic 23) B 1 f.
*Poleman 4945 (Smith Indic 194) f. 6v.
*Poleman 4946 (Smith Indic MB) IV 1 f.
*Poleman 4946 (Smith Indic MB) V 1 f.
*Poleman 4946 (Smith Indic MB) LIII f. 1.

GEOGRAPHICAL TABLE

I. Manuscripts:

*Poleman 4860 (U Penn 1895) 2 ff.

II. Tables:

The table is of geographical localities which were fairly well known in the eighteenth century and of the length of a gnomon-shadow measured in digits at noon of the equinoctial days in each locality. In column 1 I assign a serial number to each entry; in columns 2 and 3 the reading of the manuscript is recorded; and in columns 4 and 5 I give an identification where possible (sometimes they are very dubious), and the approximate latitude North to the nearest 0;10°. The spellings of the place-names, their locations within the political divisions and sub-divisions of the British Raj, and their latitudes are derived from vol. 26 of the *Imperial Gazetteer of India*, rev. ed., Oxford 1931.

1)	5;0	ujjaina (*cf.* 53,160)		Ujjain, Gwalior	23;10°
2)	5;36	(*cf.* 54)			
3)	5;6	ḍhakanapura			
4)	6;20	ḍhāṃkā		Dhak, Shahpur, Punjab	32;30°
5)	5;30	kālavāga (read 6;30)		Kalabagh, Mianwali, Punjab	33;0°
6)	7;52	kaśmīra (*cf.* 153)		Kashmir	
7)	4;25	korāgāṃvaṃ		Koregaon, Satara, Bombay	17;40°
8)	8;30	kābil		Kabul, Afghanistan	34;30°
9)	3;5	kāṃtī (read 5;5)	(?)	Kotwal, Gwalior	
10)	3;15	koṃkaṇa		Konkan, Bombay	
11)	6;10	kana ujja		Kanauj, Farrukhabad, UP	27;0°
12)	6;36	kurukṣetra (*cf.* 60)		Kurukṣetra	
13)	6;16	kapilā		Rummindei, Nepal	
14)	7;0	kedāre		Kedarnath, Garhwal, UP	30;40°
15)	2;30	kāṃcī		Conjeeveram, Chingleput, Madras	12;50°
16)	8;0	khaṃdāra (*cf.* 47)		Gandhāra	
17)	4;55	khamāca (read 5;55)	(?)	Kāmākhya, Kamrup, Assam	26;10°
18)	4;56	gaṃgāsāgara (read 5;56)	(?)	Ganga Sara, Jodhpur, Rajputana	24;50°
19)	3;0	gopā	(?)	Goa	15;30°
20)	3;45	golakuṃdā		Golconda, Atraf-i- Balda, Hyderabad	17;30°
21)	4;0	golavaṃ			
22)	5;25	gargarāṭ (*cf.* 58)		Kali Sindh River, Gwalior and Kotah	
23)	5;50	guvarāpura			
24)	5;45	gahorā			
25)	5;10	gauḍa		Bengal	
26)	5;50	gayā		Gaya, Bihar	24;50°
27)	6;4	gaṇaśuklerabhya			
28)	6;8	gorakhapura		Gorakhpur, UP	26;40°
29)	5;55	gokul (read 4;55)	(?)	Gogola, Diu	20;40°
30)	4;48	gujrāṭ		Gujarāt	
31)	4;0	bol			
32)	5;45	cāṭasta			
33)	5;0	caṃderī		Chanderi, Isagarh, Gwalior	24;40°
34)	5;30	citauḍa		Chitor, Udaipur, Rajputana	24;50°
35)	7;0	jālaṃdhara		Jullundur, Punjab	31;20°
36)	4;58	jaṃbūsara		Jambusar, Baroda	22;0°
37)	5;0	jūnagaḍa		Junagarh, Kathiawar, WI	21;30°
38)	5;36	jāro (read 5;6)	(?)	Jarod, Baroda,	22;30°
		or (read jaso)	(?)	Jaso, Bundelkhand, CI	24;30°
39)	5;47	jivanapura		Jaunpur, UP	25;50°
40)	5;10	jalālābāda (read 6;10)		Jalalabad, Ferozepore, Punjab	30;40°
41)	4;45	puṇā (*cf.* 173)		Poona, Mombay	18;30°
42)	6;0	dillī		Delhi, Punjab	28;40°
43)	3;45	viśālagaḍa		Vishalgarh, Kolhapur, Bombay	16;50°
44)	6;25	naipāl (*cf.* 99)		Nepal	
45)	7;30	lāhora		Lahore, Punjab	31;40°
46)	4;24	nāśika		Nasik, Bombay	20;0°
47)	2;20	khaṃdhāra (read 4;20)	(?)	Kandahar, Nander, Hyderabad	18;50°
48)	6;30	jharasara			
49)	3;55	tailaṃga (*cf.* 82)		Telingana	
50)	4;15	bedara (*cf.* 109)		Bidar, Hyderabad	18;0°
51)	4:0	vidyānagara		Hampi, Bellary, Madras	15;20°
52)	4;25	devagirau (*cf.* 101)			
53)	5;2	avaṃti		Ujjain, Gwalior	23;10°
54)	5;6	somanāthapāttana (*cf.* 148)		Somnath, Kathiawar, WI	20;50°
55)	5;4	nāgora (*cf.* 96)		Nagod, Bundelkhand, CI	24;30°
56)	6;24	yoginīpurapāṭalī(?)		Patiala, Punjab	30;20°
57)	5;30	citrakūṭe		Citrakut, Bundelkhand, CI	25;20°

58)	5;20	gargarāṭ		Kali Sindh River, Gwalior and Kotah	
59)	6;0	ajameri (*cf.* 157)		Ajmer, Rajputana	26;30°
60)	6;55	kurukṣetra		Kurukṣetra	
61)	5;35	ābhāsapattana (read prabhāsa)		Somnath, Kathiawar, WI,	20;50°
62)	4;51	bhṛgukache		Broach, Bombay	21;40°
63)	10;3	samarakaṃde		Samarkand	
64)	10;3	khurāsāne		Khurasan	
65)	6;50	sarasapattana			
66)	4;40	sītapure (read 5;40)		Sitapur, UP	27;40°
67)	4;57	maṃḍapācale		Mandav Hills, Kathiawar, WI	
68)	4;51	khaṃbhādrava (read 5;51)	(?)	Kambar, Larkana, Bombay	27;40°
69)		goṭeru āgaha			
70)	5;20	pīlāpāṭana	(?)	Pilu, Thar Parkar, Bombay	24;40°
71)	6;0	viṃdugariri			
72)	5;30	jagannātha (read 3;30)		Cocanada, East Godavari, Madras	17;0°
73)	6;30	sarjhara	(?)	Surajgarh, Jaipur, Rajputana	28;20°
74)	6;15	nāranola		Narnaul, Patiala, Punjab	28;0°
75)	6;36	mahal			
76)	7;0	jīda	(?)	Jind, Punjab	29;20°
77)	6;4	hiṃsāra (read 6;40)		Hissar, Punjab	29;10°
78)	6;50	pāṇipatha		Panipat, Karnal, Punjab	29;20°
79)	6;40	sonapatha		Sonepat, Rohtak, Punjab	29;0°
80)	6;35	narelā (read barelā)		Bareilly, UP	28;20°
81)	5;45	tājapura		Tajpur, Darbhanga, Bihar	25;50°
82)	4;4	tailaṃga		Telingana	
83)	6;5	dvārikā (read 5;5)		Dwarka, Kathiawar, WI	22;20°
84)	4;11	davlatābāja		Daulatabad, Aurangabad, Hyderabad	20;0°
85)	5;4	devagaḍha		Deogarh, Jhansi, UP	24;30°
86)	4;40	dadhigrāma			
87)	5;0	dāmola	(?)	Damoh, CP	23;50°
88)	5;45	damanapura (read 4;45)		Daman	20;30°
89)	5;2	dholaka		Dholka, Ahmadabad, Bombay	22;40°
90)	5;20	brahmapurī	(?)	Bramhapuri, Chanda, CP	20;30°
91)	6;30	pānapura			
92)	5;57	dhakalapura	(?)	Deulavāḍā, Sirohi, Rajputana	24;40°
93)	5;10	dhāmegaṃḍa			
94)	5;54	naravara		Narwar, Gwalior	25;40°
95)	5;2	naḍibāda		Nadiad, Kaira, Bombay	22;40°
96)	5;51	nāgora		Nagod, Bundelkhand, CI	24;30°
97)	7;40	nagarakoṭa		Nagar, Kashmir	36;20°
98)	4;47	navavadāpā (read navapāḍā)	(?)	Naupada, Ganjam, Madras	18;30°
99)	6;25	naipāla		Nepal	
100)	5;47	naimiṣāraṇya		Nimsar, Sitapur, UP	
101)		devagirī			
102)	4;20	peṣaura (read 7;20)		Peshawar, NWFP	34;0°
103)	4;30	pātharī		Pathri, Parbhani, Hyderabad	19;20°
104)	5;45	prayāga		Allahabad, UP	25;30°
105)	4;5	prakāṇa			
106)	4;30	paunī		Pauni, Bhandara, CP	20;50°
107)	5;1	pāṃḍava			
108)	5;44	puruṣottama (read 4;44)		Puri, Orissa	19;50°
109)	3;45	bedara		Bidar, Hyderabad	18;0°
110)	4;31	buḍhānapura		Burhanpur, Nimar, CP	21;20°
111)	5;30	būṃdī		Bundi, Rajputana	25;30°
112)	4;30	vānagaṃgā		Wainganga River	
113)	6;57	rajavāḍā (read 3;57)		Rajewadi, Poona, Bombay	18;20°
114)	3;30	bījāpura		Bijapur, Bombay	16;50°
115)	5;6	bhalāsā		Bhilsa, Gwalior	23;30°
116)		vadhānagara		Vadnagar, Baroda	23;50°
117)		bhikāner		Bikaner, Rajputana	28;0°
118)	4;52	paḍodara		Baroda	22;20°
119)	8;7	balatha (read balakha)	(?)	Balkh, Afghanistan	
120)	6;27	virāṭ (*cf.* 121)		Bairat, Jaipur, Rajputana	27;30°
121)	6;0	vairāṭī		Bairat, Jaipur, Rajputana	27;30°
122)	4;5	bhūcavī			
123)	4;8	bhṛgupuṣakṣe			
124)	5;30	mālavā		Mālwā	
125)	4;57	pāṃḍogaṇa	(?)	Bandhogarh, Baghelkhand, CI	23;40°
126)	6;0	mathurā		Muttra, UP	27;40°
127)	5;56	mugerī		Monghyr, Bihar	25;20°
128)	4;30	śurera			
129)	6;0	mūlatāna		Multan, Punjab	30;10°
130)	6;0	mithilā		Mithilā	
131)	5;30	ijā (*cf.* 169)	(?)	Ichapuram, Ganjam, Madras	19;10°
132)	5;28	yodhapura		Jodhpur, Rajputana	26;20°
133)	5;45	gohi			
134)	4;44	rājāpil		Rajpipla, Rewa Kantha, Bombay	21;50°
135)	0;30	rāmeśvara		Rameswaram, Ramnad, Madras	9;20°
136)	5;30	rājapura (read 6;30)	(?)	Rajpura, Patiala, Punjab	30;30°
137)	7;30	lābhapura			
138)	5;30	raṇathaṃ bhaura		Ranthambhor, Jaipur, Rajputana	26;0°

139) 5;30	lavaṃgapura			
140) 5;44	lakṣavatyā	(?)	Lachmangarh, Jaipur, Rajputana	27;50°
141) 5;45	lakhanaura		Lucknow, UP	26;50°
142) 5;25	rājamahal		Raj Mahal, Jaipur, Rajputana	26;0°
143) 5;0	siṃhā	(?)	Singpur, Vizagapatam, Madras	19;0°
144) 5;18	siṃrauja (read siraumja)		Sironj, Tonk, Rajputana	24;10°
145) 6;50	samasāvāda	(?)	Samasata, Bahawalpur, Punjab	29;20°
146) 6;3	samala		Simla, Punjab	31;0°
147) 4;28	śyāhagaḍh	(?)	Shahgarh, Saugor, CP	24;20°
148) 5;6	somanātha		Somnath, Kathiawar, WI	20;50°
149) 4;40	siṃdhupura			
150) 5;12	sāraṃgapura		Sarangpur, Dewas Junior, CI	23;40°
151) 4;47	sūrata		Surat, Bombay	21;10°
152) 5;17	sahasrāṃva (read 6;17)	(?)	Saharanpur, UP	30;0°
153) 6;46	śrīnagara		Srinagar, Kashmir	34;0°
154) 6;30	hastināpura		Merat, Meerut, UP	29;10°
155) 6;7	ayodhyā		Ajodhya, Fyzabad, UP	26;40°
156) 4;36	amadābāda		Ahmadnagar, Bombay	19;10°
157) 5;52	ajmera		Ajmer, Rajputana	26;30°
158) 5;31	ahamadābāda		Ahmadabad, Bombay	23;0°
159) 6;30	āgrā		Agra, UP	27;10°
160) 5;0	avaṃti		Ujjain, Gwalior	23;10°
161) 5;25	amarakoṭa (read 6;25)	(?)	Amber, Jaipur, Rajputana	27;0°
162) 4;36	abhānaura	(?)	Abhanpur, Raipur, CP	21;10°
163) 6;45	apasaṭilā			
164) 4;30	avaraṃgābāda		Aurangabad, Hyderabad	19;50°
165) 5;40	agnamapāṭama			
166) 5;20	iṃdura		Indore, CI	22;40°
167) 6;0	iṭāū			
168) 5;30	udayapura		Udaipur, Mewar, Rajputana	24;40°
169) 5;45	ichā	(?)	Ichapuram, Ganjam, Madras	19;10°
170) 4;6	oḍha		Orissa (Uḍra)	
171) 5;4	utkala		Orissa	
172) 6;0	brahmāvarta		Brahmāvarta	
173) 4;0	puṇeṃ		Poona, Bombay	18;30°
174) 3;55	vāīṃ		Wai, Satara, Bombay	18;0°
175) 5;0	kalakattā		Calcutta, Bengal	22;30°

Manuscripts: 4860 ff. 1–2.

MĀSAPRAVEŚAPATTRĀṆI

In this type of table it is indicated for each degree of solar motion how long it will take the Sun to travel thence 30°; the entries are in days (modulo 7) and day-fractions, so that 28 added to each entry gives the desired time. The tables are usually set up with the 12 zodiacal signs as vertical arguments, and the 30° (0° to 29°) of a sign as horizontal argument.

Manuscripts:

*Poleman 4886 (Smith 98) 1 f. Saṃ. 1861, Śaka 1727 = A.D. 1804.

*Poleman 4887 (Smith 86) 1 f.

See also *Candrārkī*, table 5.

NAKṢATRACARAṆAPRAVEŚAPĀTTRĀṆI

I. Manuscripts:

*Poleman 5181 (Smith Indic 87) 1 f.

*Poleman 5188 (Smith Indic 129 B) f. 21.

II. Tables:

The idea of these tables is to indicate the time it takes the Sun to traverse each nakṣatracaraṇa (i.e., navāṃśa) of 3;20°. Normally 4 tables would be given; the first indicates the time the Sun is in each of the first caraṇas of each of the 27 naksatras, the second indicates the time it is in the second caraṇas, and so on.

INDEX OF SIGNIFICANT PARAMETERS